SAMURAI BUSINESS

*The way of the warrior
for professionals
in the digital century*

By Joris Merks
Artwork by Thera Benjaminsen

Published by Meghan-Kiffer Press
Tampa, Florida, USA
Innovation at the Intersection of Business and Technology
ISBN 978-0-929652-21-4 LCCN 2012946314

Meghan-Kiffer Press
310 East Fern Street, Suite G
Tampa, Florida 33604 USA

Meghan-Kiffer books are available at special quantity discounts for corporate education and training use. For more information, contact Special Sales Meghan-Kiffer Press, Suite G, 310 East Fern Street, Tampa, Florida 33604 or e-mail info@mkpress.com.

MK

Meghan-Kiffer Press
Tampa, Florida, USA
Publishers of Advanced Business-Technology
Books for Competitive Advantage
www.mkpress.com

In memory of my teacher and friend
Jan Sjoerd Dotinga

1946 - 2008

Contents

精力善用

Rather than opposing a force,
accept it and then redirect it.

Develop yourself to the benefit
of the world around you.

The young boy and the Samurai

I spent the early part of my life training to be a competitive fighter. I am now thirty-five years old and I work as a research manager for Google. The journey that took me from fighter to business professional was one of luck, choice and circumstance, with many unexpected turns and lessons on the way. In the process I have had to completely transform myself.

But one part of me has remained constant. For as long as I can remember the mindset of the Samurai has been the basis for how I see the world. The Samurai principles combine power and respect and I believe that it is this combination that makes strong professionals. My background as a fighter gives me a unique insight into the daily challenges of office life that all professionals face. I would like to share this insight, my journey and my experiences; my successes, but also my losses, challenges, doubts and fears.

When I was six years old, my father suggested that I try Judo. I didn't know exactly what Judo was, but I did know the kids and the masters wore cool suits, like I'd seen in the movies. My father and I went to watch a training session. I don't remember many details from that visit, but I do remember how impressive the teacher was: a strong man with grayish hair, a wise look and a black belt hanging loosely around his waist. The black belt was worn from the many years of judo experience that were also ingrained in this man's body.

This image made a powerful impression on me, but also scared me.

The kids all looked so strong and there was something intimidating about the discipline of training. At first this put me off and I decided I didn't want to practice Judo but my father convinced me to give it a three-month trial. Judo has been an irreplaceable part of my life ever since.

As a child I watched a lot of fighting movies. They always inspired me to train more and harder. I had this image in my mind of becoming a super hero: not only strong, but also noble and wise. In fighting movies there is always a good and a bad character, and both are very talented fighters. There is one big difference though: the bad character can't handle the feeling of power that comes along with his talent. He misuses it to bully others. This behavior is mostly encouraged by his teacher, who also yearns for power and control over others. The good character also has a teacher, who watches over him like a father and helps him to develop his talents to use them with honor and integrity.

I especially remember a scene from the classic movie "*Bloodsport*." A young boy of about ten breaks into a house and tries to steal a sword from an altar. When he grabs the sword, an old Asian man walks into the room and takes the sword from him. With calm power the man explains that this is not a normal sword, it is a Samurai sword, and a Samurai sword cannot be stolen, it can only be earned. As he says this, the man suddenly draws the sword and slices the cap off the young boy's head. The old man notices that the boy didn't even blink as this happened. The boy has seen his path and wants to earn the Samurai sword. He asks the man to be his teacher, which is the start of a disciplined training program during which teacher and pupil build a strong father-son like relationship.

In my first six years of Judo I had three different teachers and I have good memories of all of them. When I was twelve years old, I met the man who would be my teacher for life. His name was Jan Sjoerd Dotinga. He was not the Asian teacher you would find in a typical fighting movie, but he was both wise and warm: a man who had been practicing multiple disciplines of martial arts throughout his entire life. He had trained under renowned masters and gathered wisdom and experience in both fighting and life through training and meditation. You could say he was an explorer of life using martial arts as the means of exploration.

I don't remember exactly how we connected or how I became his pupil for life. I remember that I found him fascinating and therefore made sure I was always last leaving the *dojo* after training. *Dojo* is the Japanese name for the building in which you practice martial arts. After training Jan always walked to the train station, which gave me the opportunity to walk along with him and ask questions.

I think that, as a teacher, he enjoyed my eagerness to learn both in training and in life. I started to accompany Jan as his assistant in training sessions. Jan taught Judo, and also Aikido, Jiu-jitsu and Kobu-do (Japanese sword fighting). I started to learn these disciplines as well and I became a practitioner of Budo. *Budo* is an aggregate name for all Japanese martial arts. It refers both to the physical and spiritual aspects of training. *Bu* means *war* or *martial* and *do* means *path* or *the way of*

誠忠義士傳

大星由良之助
良雄

万山不重君命重
一髪不輕我命經

應需　一筆菴誌

一勇齋
國芳畫

life. So *Budo* can be translated as *the way of the warrior*. At the age of sixteen I got my black belt in Judo and at the age of twenty a black belt in Aikido. While studying first physics and then science of communication at University, I became a teacher of Judo and Aikido. Jan and I spent many hours training and talking, and building a strong bond. I could go to him with any kind of question and he would help me in trying to discover the answers for myself.

When I was twenty something happened that would eventually stop me practicing martial arts completely. In the split-second that it happened I remember that the fear of never being able to fight again was the first thing that flashed through my mind. I was in the middle of a fight in our yearly Judo team competition. I'd had a great season and had won every single fight. I was in good shape and my body felt almost indestructible. Unfortunately this wasn't the case. In the rush of the fight my opponent suddenly surprised me with an attack. He unbalanced me and swept his leg into the side of my knee. I tried to escape the attack but my leg got stuck and the twisting force on my knee combined with the sweeping leg of my opponent coming in, dislocated my knee temporarily. I immediately felt that this was a serious injury and was taken to hospital.

In the months that passed, I had examinations and surgery and it was discovered that my cruciate ligament, which normally keeps the knee stable, was completely torn off. Luckily it could be replaced and after nine-month's rehabilitation I would most likely be able to practice martial arts again. At the time this happened I had been training twice a day and suddenly that was not possible anymore. This meant I had huge amounts of energy with no outlet. This was highly frustrating for me, so I bought myself a guitar and started playing guitar for six hours a day and within six months I could play some nice songs. I remember sitting in a bar with my injured leg on a chair, talking to my teacher. I told him about the guitar and played him some songs. He smiled and said: "you are a true Samurai." I know he meant it from the bottom of his heart and I still carry this moment like a treasure.

After nine months of intensive rehabilitation I discovered my knee was not stabilizing. Whenever I tried doing Judo again, or anything close to it, my knee dislocated in painful ways. Each time this happened, I lost my ability to walk for a few weeks and had to start the rehabilitation process again. Over a period of about ten years I had five surgeries and have been more or less continuously in rehabilitation. After about five years I accepted that practicing Judo was not an option anymore. However, I discovered Brazilian Jiu-jitsu, which was similar to Judo but more focused on ground fighting. There was no referee coming in between the fighters and you could not decide the fight by throwing someone as in Judo, so you always had to win by fighting on the ground. If I was fighting on the ground, I did not have to stand on my knee and I could still enter competitions. In 2002 I became European Champion and Open German Champion in Brazilian Jiu-jitsu. Winning those titles was an important step in getting to the point where I could let go of the desire to continue practicing martial arts.

Even though I focused on ground fighting, I knew I was still taking a big risk with my knee. You can never fully control what happens in a fight. I gradually came to accept that my knee would never heal again but that I still had to walk on it for the rest of my life. This meant I had to stop practicing martial arts and that I would never be able to do sports that require running and turning, like soccer and tennis.

This stupid two-centimeter ligament had in one second swiped away everything I had trained for since I was six years old. My whole being was that of a fighter but my body wasn't capable of supporting that self image anymore. Throughout my whole life I had been able to solve problems by finding creative solutions or by just trying harder. In this case, the harder I worked at solving the problem, the more I damaged my knee. I just had to accept that this was a problem that could not be solved. This brought about a personal crisis, for I had to completely re-invent who I was. My brain still thought I could make spectacular fighting moves and that I could run and jump, but my body couldn't do that anymore. I constantly had to remind myself of that fact.

Even though I was not training in martial arts anymore, I stayed in close contact with my teacher. We met during weekends and walked or cycled through the forest spending hours talking. He helped me to disconnect from my self-image as a fighter and replace that with something else. That "something else" was being a Samurai. For the true way of the warrior extends beyond the physical practice of martial arts. It is a continuous curiosity and desire to understand life and yourself. This curiosity had always been there, I just had to learn to separate it from physical exercise and by doing so I learned to understand the true values of the Samurai.

If you have seen any classic fighting movies you will know that in the end the master dies. My master, and friend, Jan also passed away. I remember getting the phone call from his son while I was sitting in the office of the research agency I worked for at the time. Jan had had a sudden heart attack the evening before and was most likely not going to make it. I couldn't really believe it. Jan was so healthy and always looked so full of positive energy. At the same time I felt an immediate gratitude for the fact that I had been given the privilege of meeting this man, of having him as my friend and teacher and for spending so many hours talking and training with him. I felt grateful for his lessons, wisdom and unconditional friendship. I knew I would carry those along wherever life would take me.

Along with Jan's children I was one of the pall-bearers at his funeral. They are still like a family to me. Jan's wife gave me the Samurai sword that he used to train with. I remembered the master in "*Bloodsport*" saying: "a Samurai sword cannot be stolen, it can only be earned." I always create a special place in my home for this sword. It is one of three objects that help me to cherish the memories of my teacher. The second is a meditation bench that Jan gave me, because I wasn't able to kneel and greet as is normal during Judo classes after my knee injury. The third is the book of Musashi that Jan also gave me that reminds me of his lessons.

In this book "*Samurai Business*" I share the lessons of my teacher and the wisdom of the Samurai. Business is in many ways competition and competition requires skill, strategy, drive, creativity, endurance, patience and knowledge, of yourself and your competitor. These are also the skills you need as a fighter. I hope this book inspires you to become the best professional you can be, without the need for office politics. I find the approach valuable because the aim is for harmony without weakness, and understanding of both self and others. I hope the teachings of the Samurai provide a refreshing way of looking at the contemporary business world and how you function in that world.

You could say this book is similar to Sun Tzu's "*The Art of War*," but modernized for the digital age. In this digital age, businesses need to be more transparent and to be as fast and flexible as their changing environment. This means that you, as a professional also need to have those skills. I believe the increased transparency brought by the Internet, has made the role of the individual more important. The majority of this book therefore focuses on strategy and self-development for the individual professional. This book is about being focused in the middle of an endless sea of stimuli, about staying rooted in who you are and at the same time staying open and connected to your environment – to your colleagues, consumers, clients and competitors.

You will recognize many of the lessons from experience or maybe from having heard them before. I feel however that organizing them into one coherent metaphor of the Samurai in business makes those lessons more powerful and inspiring. I hope you will enjoy reading this book as much as I enjoyed writing it and that it helps you in making your work and life a playground of experimentation and personal development.

A fighter becoming a business man

In the years when I was gradually learning to accept my permanent injury and was mentally preparing to end my life as a fighter, I started my professional career. I first worked for a media agency, then for an advertising agency and finally became an expert in media, advertising and marketing research. This brought me to my current role as research manager for Google.

Throughout the startup of my career, by working for various companies I have become accustomed to working with normal people (to me these are people who are not fighters) and I have learned the unwritten rules that all work spaces have. I feel happy in my work and year by year I settle more deeply into an acceptance of no longer being able to fight. It has taken me a few years to get to this point though.

I vividly remember how I felt when I first started working in an office. I had grown up in the world of martial arts; I knew the rules and understood the people. In my eyes it was a peaceful world of disciplined people seeking self-development and truth. The office on the other hand felt almost like a war zone to me. I was surprised by how aggressive people could be when chasing their personal objectives in a professional environment. The worst part was that it was not a form of open battle. The way people fight in offices is not transparent at all, however everyone seemed to find this normal and seemed to know the rules except me. It was ironic that I, a fighter, felt unsafe in this normal world, which in my eyes was filled with aggression.

For example, coming from a sports background it was natural to me to be eager to learn, so I would take on any project for any fee as long as it would help me in my development and in gaining experience. I had no limits as to how hard I was willing to work. I expected that my co-workers and managers would appreciate this attitude and would help me to develop so that we could achieve more together.

I never imagined that an eager new person entering a company might be considered a threat by the people already working in the office, or that the people who were supposed to be my colleagues might actually be my opponents. I never expected that those people would take my willingness to work on any project and use that enthusiasm to give me all the most annoying projects that no one else wanted to work on or which everyone else knew had a high risk of failing. I never expected people to give misinformation with a friendly smile or consciously withhold information in the hope of undermining my efforts. I never expected that a manager might thrive by making workers feel they are performing badly, while at the same time taking credit for their accomplishments.

I have now discovered that many of these behaviors are more or less standard in almost any office. Some companies are even dependent on some of these mechanisms in their business model. Young people are always eager to perform well and are often willing to accept a low starting salary. They have little clue about the power structures in the office, so they are easy targets. If young people starting out in companies refused to accept this combination of heavy workload and low pay some companies would not be able to maintain their margins.

Even stranger: the more you accept the project overload, and the more you prove your unconditional willingness to work, the lower your reward. People don't tend to care about how many things in total you are doing. They just want you to finish the *one* project that *they* are interested in. The chance of a supervisor or manager recognizing all the many things you are accomplishing is very low unless you start making objections about your workload, but if you have unconditional commitment and are used to training extremely hard you are unlikely to object to a heavy workload.

It took me quite a while to discover that I got more appreciation for my work if I refused to do more than a certain number of projects or refused to accept projects that I didn't want to work on. If you never say "no" your effort becomes like water: it is always there and therefore loses its value. As a result people in offices have created the habit of always highlighting how busy they are. As far as I am concerned, if you have time to demonstrate your busyness, or to talk down your colleagues, you aren't busy *enough*.

Ironically the unconditional willingness to learn and enthusiasm for work also has an inverse relation to being paid well. I remember having a job interview and being asked how much I wanted to earn. I responded by saying I did not care about my salary as long as I would be able to gain experience. I said I felt confident that I would prove myself, which hopefully would result in higher salary later. This of course resulted in a low salary and I settled for that without any second thoughts.

I have now discovered that in most companies you don't get a substantial pay raise until you demand it. In a lot of cases managers will offer a sudden big raise when you find another job. I have always thought this strange as it is basically a reward for disloyalty. I wonder why they aren't willing to offer the same pay raise while someone is still a hundred percent committed to the company. An offer of increased salary after handing in my resignation has never made me change my mind.

I remember in the first years of my career, office politics made me feel like a blind man entering a battle arena. I would feel attacked, but I could not see exactly what was happening. My body always reacted as though I was in a fighting situation. My brain shot small and invisible fight reflexes through my body, which I constantly suppressed by reminding myself this was not a fight. There have been some occasions where I became so angry about a supervisor betraying my trust and dedication that I raised my voice in anger. Generally however, I managed to control my emotions while looking for ways to understand what was happening and for new ways to cope with the situation. I have observed myself and others like this for years now and have developed a more sensitive eye for political behavior. I still don't like it, but I generally know how to make sure my work is not affected by it.

As my career progressed I felt my old fighting skills coming back in a new form and context. As these skills returned and took on new meaning, the feeling of losing my previous life as a fighter receded. Strategic mental interactions between people and companies have many similarities with physical interactions between fighters. The endless complexity of human interaction in the context of fighting has always fascinated me. The complexity of people interacting with each other or with brands invokes a similar fascination. In my work as researcher I try to quantify irrational behavior and emotions in order to understand them. Once behaviors and emotions have been transformed into numbers, I bring those numbers to life again in trainings and presentations in conferences to help people understand the thoughts and feelings of their digital century consumers.

I have worked in marketing and advertising research for about fifteen years; I have probably been to a thousand business meetings and hundreds of presentations. I have found my own Western approach for applying ancient Japanese words of wisdom and utilizing the lessons of my teacher. In the end it was just a matter of learning to understand the language and unwritten rules of the office environment. When I feel myself using my fighting skills beyond the world of physical fighting, I can often hear the voice of my teacher as an echo in my mind: "you are a true Samurai."

The Samurai: a quest for self-development

Samurai is the term for the military nobility of pre-industrial Japan. The term can be translated as *those who serve in close attendance to the nobility*. Many Samurai were famous for both their skill with the pen and with the sword, or the *bun* and the *bu*, the harmony of fighting and learning. By the thirteenth century, Zen Buddhism had begun to shape Samurai standards of conduct, particularly in relation to overcoming the fear of death and killing.

Emperor Meiji abolished the Samurai's right to be the only armed force in favor of a more modern Western-style army in 1873. Samurai were no longer allowed to wear a katana (Samurai sword) in public. The Imperial Japanese Armies were conscripted, but many Samurai volunteered as soldiers and many advanced to be trained as officers, as they were highly motivated, disciplined and exceptionally well trained.

In the first chapter I mentioned that my teacher gave me a book about the life of a Samurai called Musashi. Miyamoto Musashi lived from 1584 to 1645 and achieved renown through the stories of his excellent swordsmanship in numerous duels, from a very young age. Musashi wrote "*The Book of Five Rings*," a book on strategy, tactics and philosophy that is still studied today. He created and perfected a two-sword technique using both a large sword and a *companion sword* and was an expert in throwing weapons. He frequently threw his short sword in combat with great accuracy. Musashi is widely considered one of the greatest Samurai warriors of all time.

It is said that Musashi studied at the Yoshioka-ryū school, which was also said to be a school Musashi defeated single-handed in his later years. He had formal training from either his father or uncle from around the age of seven. Musashi left his village at the age of fifteen. He travelled extensively in Japan in a warrior pilgrimage, during which he honed his skills with duels. He often used a wooden sword while fighting opponents armed with real swords. In this time the death of one of the fighters was not the aim of an engagement unless both parties agreed, but it is known that Musashi's mastery was so great that he did not care which weapon his foe used. By conservative estimate Musashi fought over sixty duels and was never defeated, although this most likely doesn't include deaths by his hand in major battles.

During Musashi's travels he became dissatisfied because he had never met an opponent who could match him. He also began to feel that there must be more to learn. At about twenty-seven, Musashi began practicing Zen meditation. At some point he met a wise old man who practiced a diversity of arts. In conversation Musashi realized that this man was more developed than himself even though the old man was not even close to being capable of defeating him in battle. This is when Musashi started practicing the arts. He made Zen brush paintings, calligraphy and sculpted in wood and metal; records also show that he had architectural skills. In "*The Book of Five Rings*" he emphasizes that Samurai should understand professions other than fighting.

When he was about thirty, Musashi fought his most famous duel with Sasaki Kojirō, who was known as *The Demon of the Western*

Provinces. Musashi came late and unkempt to the appointed place, the island of Funajima. The duel was short. Musashi killed his opponent with a wooden sword that legend says he had carved from an oar used on the boat that carried him to the island. Musashi's late arrival is controversial. Sasaki's outraged supporters thought it was dishonorable and disrespectful, while Musashi's supporters thought it was a fair way to unnerve his opponent. Some believe that Musashi timed the hour of his arrival to match the turning of the tide: the tide carried him to the island and after his victory the turning of the tide assisted his flight from Sasaki's vengeful allies. Another theory suggests that he timed the fight so that the sun would dazzle Sasaki.

The 970 page long story of the life of Musashi demonstrates that the true way of the Samurai is about seeking continuous development rather than victory. Musashi's pilgrimage starts with his physical strength driving him to challenge stronger opponents over time. As he does so, he develops tactical combat skills allowing him to fight groups of people. In the end, Musashi discovers that fighting no longer delivers enough angles for self-development, which leads him to explore new art forms. I think this is an important lesson for professionals. If your focus is on outcome and success, you will be tempted to limit yourself to your existing skill set. If your focus is on development, you will look beyond the things you can do now and develop further as a professional and as a person in the long-term.

A new era for businesses with a purpose

People often relate business and war to each other. In many cases a link is made to Sun Tzu's "*The Art of War.*" Unfortunately those comparisons often elevate the concept of *war* over the *art*. I believe the concepts of fighting and business or being a warrior versus being a strong business person are often misunderstood. I am sure there is a good reason for the fact that Sun Tzu chose "*The **Art** of War*" as his title rather than for instance "*The **Thrill** of War*" or just "*How to Destroy Your Enemies.*" "*The **Art** of War*" relates to the ancient Samurai: warriors who were more than just fighters. In the last chapter I revealed the continuous quest for understanding and self-development that is inherent to the way of the Samurai. Growth and improvement are more important than victory and outcome. As a result, these warriors became highly skilled and won more battles. The very fact that Samurai did not care about winning or losing made them more successful.

I believe this works in the same way for companies. If companies become too preoccupied with earning money, they risk forgetting what made them successful. The heart of success is building a great product or service that is appreciated by consumers and customers. Yes, businesses should be financially successful, but their financial success is not their reason for existence. Mark Zuckerberg illustrated the importance of this notion when Facebook went public. In his announcement he stated: "Going public is an important milestone in our history. But here's the thing: our mission isn't to be a public company. Our mission is to make the world more open and connected."

Successful companies have clarity about their reason for existence beyond earning money. For Google the mission is: "To organize the world's information and make it universally accessible and useful."

Google's founders, Larry and Sergey, have diligently stuck to this vision. They built their search engine with the vision of providing people with the fastest and most accurate answer possible when they were searching for information on the Internet. Until they found a way to make search advertising relevant and non-intrusive they refused to have advertising on the search page. Advertising on the Google search page is managed by an algorithm that is based on an auction and at the same time embeds a diversity of metrics quantifying relevance for the user. No advertiser can buy their way to the top of the advertising results without being relevant and the advertisements are always clearly separated from the organic search results.

Apple is another company that states their purpose clearly: "Apple is committed to bringing the best personal computing experience to students, educators, creative professionals and consumers around the world through its innovative hardware, software and Internet offerings."

This mission states that the combination of hardware, software and Internet offerings should result in a great computing experience, which is exactly what Apple achieved. They focused on building an easy to use and attractive user interface with their software, combined that with great hardware and built huge success linking the two to their iTunes store.

Companies need to continually remind themselves what their reason for existence is beyond earning money. If they stick to this reason for existence and manage to evolve along with their evolving consumer that will result in financial success. Focus on development is the source of success; financial results are the outcome, not the source. The hardest part is staying in tune with the evolving environment and consumer. The rise of the Internet is a development that has clearly changed the dynamics for many products and services. The traditional music industry tried to fight music downloads for years and Apple was the company that proved that consumers were willing to pay for music downloads; the format just had to be accessible and attractive. Staying in tune with the consumer and adapting to a changing environment, while at the same time staying close to your core values, are fundamental skills for any company. This ability becomes more important as the pace of change accelerates. Similarly, the essence of effective fighting lies in the ability to adjust to rapidly changing situations and within that to finding opportunities to strike, without losing your own balance.

The digital reset for ethical standards

The rise of the Internet is a development that has had so much impact on world structures and dynamics that it can be compared with the invention of the printed word. Printing created the ability to store and reproduce knowledge and information, so it could be widely shared. This sharing of information resulted in an increased learning curve for billions of people. Suddenly, building on existing knowledge became much easier.

The Internet offered another giant leap in accessing knowledge and information giving anyone the ability to share and consume information in a highly flexible format. It has never been easier to inform yourself on any topic you can imagine. Encyclopedias in book form have basically become pieces of furniture with a high nostalgia value.

It is impossible for governments or companies to fully control the content people have access to on the Internet. This means transparency is greater than ever. Companies can no longer get away with malpractice or low quality products. A single consumer review shared by millions can easily put an end to undesirable behavior driven by low business ethics. During the fall of the Egyptian government we saw how individuals can create power by connecting to each other using the Internet. Social networks played an important role in helping civilians to organize themselves and put pressure on the government to step down. The magnitude of interconnections between people has become so much bigger that there is a dramatic shift of power towards the connected individual.

It is now much harder for companies to be financially successful while having low business ethics and a limited purpose beyond earning money. Businesses need to learn to serve their consumers and clients better. They need to develop a new type of sensitivity to changing consumer needs. They need to serve those creatively and flexibly. Too much emphasis on power and control undermine the adaptive power of businesses and therefore their learning curve. Young, agile businesses will then learn to read the strengths and weaknesses of power driven traditional businesses and find creative ways to serve consumers better, creating more value for less money.

In fighting strategy you can find similar dynamics. There are basically two types of fighters. The first type is tactically and physically stronger. This fighter tends to be predictable, executing a consistent strategy with a lot of force. When up against this type of fighter you tend to feel under pressure and it is often wise not to directly fight the force, for it will drain your own energy over time. In this situation your best option may be to discover your opponent's weaknesses. When you target a fundamental weakness, you may also discover that these fighters lack the adaptive skills needed to retaliate against such an attack. The second type of fighter is more versatile and agile and relies less on power or fixed structures. There is still a clear strategy, but it is less predictable. This fighter tends to use less force in creating opportunities, but tends to lure the opponent into traps and responds to openings instinctively. With this fighter it is harder to find consistent weaknesses, because he keeps changing flexibly over time and invents many actions on the spot during the fight. Both strategies can be successful, however there are clear trade-offs. If you constantly put your opponent under pressure, it is harder to feel their balance and movements and adjust to those. If you have a more agile style, you tend to take more risks, which might result in exposure to counter-attacks.

Because the business environment is changing faster than ever, companies need more professionals who can listen to their environment and adapt to it creatively. Professionals should therefore rely less on politics and more on cooperation and self-development. For teams as well as individuals, this will lead to the highest possible growth curve in the long-term and above all it will result in a more enjoyable working environment and better products and services. Professionals and businesses capable of combining high ethical standards with the ability to creatively serve evolving consumers have the future in their hands.

Integrity requires true professionals

The people are the company so integrity in business requires personal integrity. I believe companies need professionals who can put their immediate demands for success and money aside as these lead to short-term thinking and tactical behavior, rather than the creation of sustainable value. Like the Samurai, contemporary professionals should have continuous self-development as their ultimate goal. Clients and end users will then be better served, with financial results as a logical consequence for both the professional and the company. Everyone serves clients to a certain extent. A client can be a person within your own company who is dependent on your work. In many ways you can say a manager needs to serve his or her team members and consider them to be clients. A client can even be a child at home if you are a full-time mother or father. The reward for doing well as a full-time parent of course is not financial in nature. The reward in that case is seeing your child grow up and become a well-balanced adult, which might be worth a lot more than money.

Serving your client does not mean you need to be nice all the time and do everything the client asks. You should serve your client while staying balanced as an individual and while keeping your business financially healthy. Being too nice can negatively affect both you and the financial health of the business you are in. For example, if you are in direct personal contact with your client and you never say "no", that client may start to demand more of you than you can deliver within an acceptable time frame or within the terms of the financial agreement in place. That may put you under so much pressure that it negatively affects your ability to serve that client or other clients in your portfolio. Managing expectations and asking for a mutually fair price is an important part of serving your client. Pricing your products or services too low is not beneficial to your clients. In the long-term, if pricing is structurally too low it will result in lower margins, which means the company has less power to continue innovating and to serve the end user or client.

So in any working environment, there is always a give and take balance between self-development and serving your end user. Finding the balance is harder when there are conflicts of interest in your working environment. The reality is that almost every professional works in situations where there are at least some conflicts of interest. Those can be both internal and external. The most obvious external conflicting interests are the objectives of your competitors. In some cases clients can also introduce conflicting interests by having unreasonable expectations of your service, product or pricing. You could say the emergence of illegal music downloads is a form of an unreasonable client expectation. Some people are not willing to pay for music anymore when they can download it for free, however if the music industry doesn't earn enough money there will eventually be very little music to listen to. So there is a conflicting objective between companies and end users and it is fair for music companies to take action on this.

There are however multiple ways to approach such a problem.

You can exert power by filing lawsuits against download services or you can adapt to your environment by using the latest techniques to better serve your end users. The latest music download techniques and online streaming allow for music to be distributed at much lower cost than before. This means legal access to music for a lower price should be possible. Online music services as delivered by Apple and Spotify prove that this can be done. End users and companies can agree on the value of a product as long as professionals focus on continuous development.

Internal conflicts of interest can often be a lot harder to identify and tackle than external ones. These can come from colleagues with a hidden agenda or from departments within your company that compete for the same budgets as yours. Since the emergence of the Internet, many companies have a split between departments for online marketing and offline marketing. Both should be serving the same end user, however, since they often have separate targets, the departments tend to compete internally which often results in political behavior.

A final example of a situation full of politics and conflict of interest is company reorganization where the number of jobs is being reduced. When people feel their jobs may be threatened the worst office politics erupt. The manager who needs to cut job roles always has to choose between two or more evils. Firing people you work with and who may have become your friends is never nice. Not firing people or firing the wrong people will threaten the survival of the company and thereby the jobs of even more people. As a professional, you are to a certain extent a warrior and warriors sometimes need to exert power. Some people think and say that only people with a bad conscience can make it to the top. I do not believe this is true. I believe you will not make it if you lack the capacity to make difficult decisions and be tough when required, but you will at the same time be less successful in the long run if you cannot balance the use of power with a fundamental sincerity. I believe this has always been true, but it is becoming even more so in this era of transparency fueled by the growth of the Internet.

You can fight with integrity no matter how strong the conflicts of interest are and integrity does not necessarily compromise your ability to stand your ground. Fighters in a duel always greet one another with respect before the fight starts. Both know the battle will be sharp, but that is no reason to despise one another. The art is in remaining respectful and doing what is right in the long-term and under pressure.

While integrity is important, every professional needs to train for tactical battle. Sooner or later every professional encounters a political environment that requires more than just the sincere will of personal development and service to your end user or client. As a competitive fighter you always need to be aware that your opponent might do something that is against the rules of the game and that he may get away with it. That doesn't mean you should cheat too, but you should make sure you are not harmed or distracted by it. Because tactics are an inevitable part of business there is a later chapter on coping with tactical resistance.

In this chapter though, I want to stress that the heart of all financial and professional success is and will remain, your sincere will to be the best professional you can be and the will to serve your end users and clients to the best of your ability. Keep these objectives firmly engrained in your mind. Tactics can be learnt but true integrity needs to be lived and you can only acquire it through deep personal transformation (or group transformation if we are talking about company culture). You cannot act *as if* you have integrity. True integrity lasts; fake integrity fades away under pressure, reducing the scope for successfully faking it again in the future.

Prisoners in office politics

If you ask professionals, the majority of them will say they dislike office politics. But in the end most people still take part. I believe this is because office politics tends to have the nature of a prisoner's dilemma. The problem is that political behavior often pays in the short-term but I am convinced that this isn't sustainable in the long-term (despite some exceptions). I would like to illustrate the limitations of political behavior with an example from my experience in teaching Judo.

When children are in the early stages of learning Judo, they often discover that the easiest way to throw an opponent onto his back is to grab both legs and throw him over. Most other techniques are harder to learn and therefore, in the short-term, are less effective for most children. If the teacher doesn't intervene children will keep applying the simplest attack because it works for them. In most cases they won't develop a diversity of techniques, nor will they develop the skill of reading their opponent and adapting to the situation. After a few years, other children will have developed a defense against simple attacks and will also have a more diverse arsenal of techniques and tactics. It is almost impossible to make up for this lack of technical development at a later stage, because it is gradually built up through years of experimenting, evaluating and improving. As a result, children who look for the easy way out in the short-term gradually start losing matches more often as they get older and their learning curve flattens out.

I believe office politics work in similar ways. Time you spend on promoting yourself or putting others down cannot be spent on self-development. So people who spend too much time on political actions, risk cannibalizing their long-term learning curve. As mentioned in the previous chapter, this is a matter of balance. Almost no one can get away with not spending time on office politics at all. You just have to evaluate critically, in the light of long-term objectives, *how much* time you spend on it. People who are too focused on short-term success and money will tend to spend too much time on politics, which means others have a good chance of overtaking them by keeping an eye on what matters for the long-term: developing yourself to the maximum of your potential and serving your end user or clients creatively.

Too much political behavior not only undermines personal development, it also makes you lose credit with the people around you (or at least you don't earn additional credit). For example: if you are a sales person selling rubbish to reach your target and bonus in one year, clients are less likely to buy from you in the future. This will make it harder to reach future targets. Even if you change job, this loss of credit will follow you. If you work long enough in a certain field of expertise, you will see the same people popping up in different job roles and each time that happens they will remember the last interactions they had with you. In a company versus consumer relationship, this mechanism works the same way: if you sell a product which delivers disappointing performance in relation to the price you asked, people will not buy your product again and they will not recommend you to others, thereby reducing your future selling capacity.

The same thing happens with colleagues. Each time you fail to recognize someone else's success or even claim it to be yours, each time you talk someone else down and they discover it, each time you make someone else do an annoying job without being honest about it and each time you ask for someone's help without returning a favor, people will remember it. Again, if you work for long enough in a certain field, you will see the same people popping up and they will think about the last time they came into contact with you. I have found myself in situations where I have been in a position to make hiring and purchasing decisions in relation to people who had previously worked above me and did not behave in very ethical ways. That kind of knowledge affects your decisions. You don't want someone to be your vendor if you know they can't be trusted and you don't want to hire someone untrustworthy as a member of your team.

People who rely too much on political behavior tend to leave a trail behind them of dissatisfied colleagues, clients, end users or consumers. This trail of damage gets larger over time. On the other hand, people who rely more on self-development and using their skills to serve their end user or client, build up a similar trail of positive credits. This trail also grows over time and will make it easier to grow in success. If a client knows you are a sales person who only sells quality products you will build up a certain amount of trust. If you meet the same person in a different situation, he or she will be more likely to buy your services again or maybe hire you. If colleagues know you acknowledge their successes and return favors whenever you can, they are more likely to

help you over time.

Again, this is something you cannot fake. You might succeed in managing that in the short-term, but faked sincerity will eventually result in behavior that is not consistent and people will notice it. The later they notice it, the more they will feel betrayed. So the longer you get away with it the worse the damage.

If you have a senior position in a hierarchy, acting with integrity towards the people who work for you gets harder. You have more power so you can easily get things done by demanding them, rather than by motivating people. Demanding certain actions can bring you faster results in the short-term. *Inspiring* people to do the right thing can sometimes take more time or effort. However each time you use the difference in hierarchy to force someone to do something, you will lose that person's respect. Each time you inspire someone to do something you will *earn* credit. If it is not needed, don't use the power you have. The use of power is a necessary for every fighter but you always need to be selective about when you apply that power, even if you are stronger than your opponent. If you use too much power in the early stages of a fight, a tactical opponent will patiently await the opportunity to score after fatigue sets in.

Managers can help their employees by making it easier to disconnect from office politics. They should make sure political behavior has minimal pay-off by having a clear vision of sustainable success and by encouraging team spirit. Sustainable success can be encouraged by not only having targets and bonuses for client spend but also for client satisfaction. Splitting bonuses into a team component and a personal component can encourage team spirit.

You could say offices have a need for heroes: people who can set an example and who thereby reduce the level of politics around them, so everyone can focus more on creation of value. This is also true in fighting. Each style of fighting has its 'street fighters' and its 'gentlemen'. Both can do well, but the gentleman fighter is most likely to develop into a hero. Anton Geesink was the first ever Western Judo competitor to defeat the Japanese by becoming World Champion in 1961. For Japanese people this was a moment of shame. When Anton won the finals one of his fans was so outrageously happy that he wanted to run onto the mat to celebrate with Anton. This would have been a terrible sign of disrespect in the eyes of the Japanese, right at the moment when they had experienced shame due to this loss. Despite the rush of the victory, within a split-second, Anton had calmly but clearly gestured that his fan should stay off the mat. He thereby showed respect to the Japanese people who invented Judo and made it what it is today. Nowadays Judo teachers proudly tell this story to their students and Anton thereby still shapes the ethical behavior of millions of Judo practitioners today.

Mark Huizinga is another hero champion in the Judo world. He is a true prodigy both tactically and technically. His creative style has inspired and entertained Judo lovers for many years. His fighting spirit is uncompromised, but he always remains a gentleman. When Mark became an Olympic champion it was a cause of celebration for many.

It felt like good had conquered evil and his screams of joy showed nothing but gratitude. There was no moment in which Mark showed any sign of arrogance due to his success. He is a truly inspiring hero.

So even though knowing how to cope with office politics is an important skill, don't overdo it. Think critically, in the context of longer-term objectives, about whether you want to spend your time on tactics or on true development and value creation. Integrity can be a powerful driving force when applied consistently and when the right tactical skills are in place. When your efforts in self-development result in success, be grateful and stay focused on further development. If you are victorious in situations of political battle, remain respectful and focused on further development.

Aikido

Aikido is a Japanese martial art developed by Morihei Ueshiba (1883 – 1969). The word Aikido consists of three parts:
- ai – joining, unifying, combining, fit
- ki – spirit, energy, mood, morale
- dō – way, path

Aikido can thus be translated as *the Way of unifying life energy* or as *the Way of harmonious spirit*. The term *aiki* in Japanese is often combined with other characters creating concepts like *unite/combine/ join together* and *mutual agreement*. You could say Aikido is a *Way of combining forces*. The term *aiki* refers to the martial arts principle or tactic of blending with an attacker's movements for the purpose of controlling their actions with minimal effort. One applies *aiki* by understanding the rhythm and intent of the attacker to find the optimal position and timing to apply a counter-technique. This means you need to connect sensitively to your opponent in order to succeed in your defense. Ueshiba's goal was to create an art that practitioners could use to defend themselves, while also protecting their attacker from injury.

Aikido can be seen as the grammar of all martial arts, because it makes the basic principles of fighting explicit. Typical of Aikido are fluent, curved movements that make it attractive to watch. Aikido is generally practiced in a non-competitive form. It derives much of its technical structure from the art of swordsmanship.

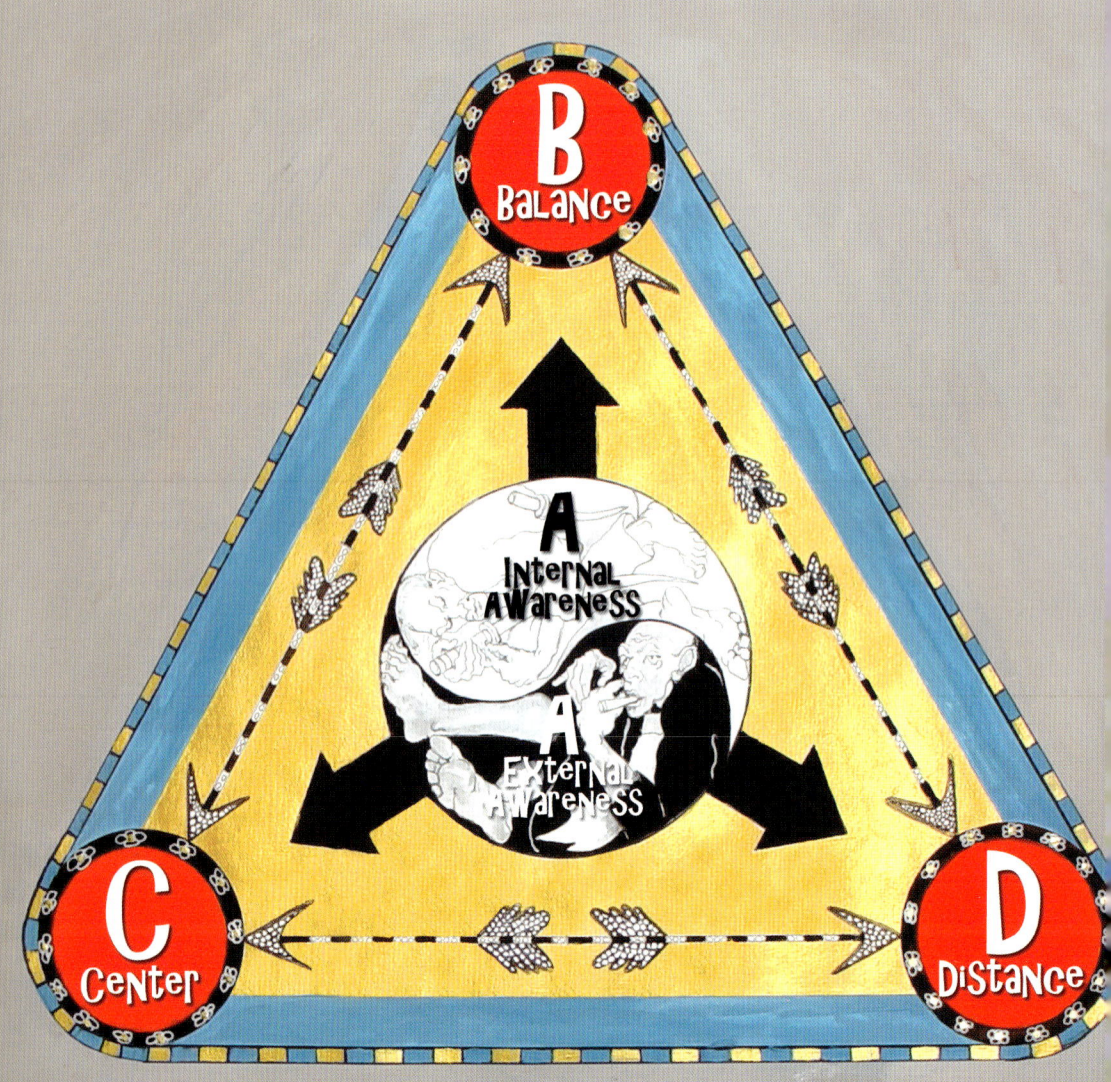

Basic principles of fighting effectively

There are thousands of styles of martial arts with their origins in different countries but they all have the same overarching principles that determine the quality of a fighter. Those principles are:

1. **A**wareness (internal and external)
2. **B**alance
3. **C**enter control
4. **D**istance control

A fighter who can harmoniously control all of these four principles at the same time will be capable of handling threats in the environment effectively and efficiently. I will walk through each principle in the context of fighting and also touch upon its relevance for the business context.

Awareness

Everything starts with awareness. Awareness is the ability to know what to pay attention to *and* what *not* to pay attention to. In a fight it is important to see attacks coming as early as possible so you should monitor all your opponent's movements. However, those movements will contain a lot of intended and unintended distractions. Not every movement is a potential attack. The ability to know which movements to monitor and which to ignore is very important. The same is true in business. There is always too much information and the ability to select what is important and what is not, is crucial in being successful as a business professional. Many things that seem to be a threat in the short-term are not harmful in the long-term, and vice-versa. It is therefore important to develop the skill of seeing distant things as if they were close and of taking a distanced view of close things.

Without awareness, there is no alignment with the constantly changing environment. On top of that, without awareness there is no self evaluation of performance and no learning curve. Awareness has an internal and an external component. A fighter uses the internal component to assess his own strengths and weaknesses, while simultaneously the external component tracks those of the opponent. Like a fighter, a professional should be capable of simultaneous internal and external tracking. Listen to others, but don't get trapped in their opinions. Serve others, but not in a way that harms the fundamentals of who you are. Know the strengths and weaknesses of your company in relation to your competitors and your environment. Finally, you should know your *personal* strengths and weaknesses in relation to the colleagues around you. Since all variables constantly change both internally and externally, awareness is an 'always on' activity.

Balance
In fighting, balance can be mental or physical in nature. Using the principle of internal and external awareness, balance applies to both you and your opponent. You can only unbalance an opponent if you are

balanced. The more you are rooted in your body, the more you will be able to feel early signs of balance disturbance, which you can automatically correct.

As a professional, the more you are rooted in who you are, the better you can capture the early signs of disturbance of your mental or physical wellbeing. As a company, the more you are in tune with the people working for you, the sooner you can detect signs of unhappiness, work overload, inefficiencies or any other problem.

For individuals the motivation for correction of disturbances often comes from self-respect. Self-doubt can significantly harm your ability to maintain your balance under pressure. I, for instance, discovered that my desire to be liked or my fear of making mistakes could unbalance me. By pressing the right buttons people could persuade me to take on more work, even though I knew my workload was already too high. If you have weaknesses like this, political people will find them and unbalance you before you consciously feel it. Someone might approach you by saying "We have an urgent problem and you are the only one who can solve it." If you have a need to be liked or acknowledged, this remark might touch your ego and persuade you to take on work that you should turn down.

Balance is related to risk taking. People who take too many risks can be compared with a fighter trying to unbalance an opponent, while forgetting to maintain his own balance, thereby exposing himself to a counter attack. People who take too *little* risk can be compared with a fighter who remains well balanced, *but forgets* to challenge the

balance of the opponent, thereby allowing him to take the initiative. So you should take risks, but monitor the potential for these backfiring. Make sure you feel comfortable, but without staying in your comfort zone. Larry Page, co-founder of Google, formulated this inspiringly as "uncomfortably exciting."

Center control

Imagine yourself opening a jam-jar. When opening a tight lid you will automatically bring your hands to your belly. This is the center of your power, where you generate most strength. A fighter should always arrange his body in a way that allows him to generate power from the center. You don't make a powerful punch sideways and a kick can only be powerful if the center of your body is powered firmly by the supporting leg. At the same time, the opponent should be positioned off-center to reduce his force. Judo players often do this by pulling opponents down by the neck, making them bend over. In sports like (kick)boxing, Karate or Taekwondo center control of opponents can be challenged by moving in a circle around them. If an opponent doesn't follow, he will move off-center.

It is important to simultaneously challenge the center power of the opponent, and stay centered yourself. This is also true of the business situation. The art of brand management for instance, is making innovations that are refreshing and new without losing connection with the core being of your brand. Companies or professionals with too much focus on external awareness tend to become copycats. They get scared if they see competitors being successful with innovations and start copying them. Over time they gradually move away from the essence of the sustainable competitive advantage of their own company. As a result, in many product categories consumers perceive very little differences between brands; this results in reduced loyalty and more pressure on price, so margins become lower.

Companies with too much focus on *internal* awareness risk getting stuck in fixed routines, and losing the freshness of brands and products. If you apply center control to the interaction between people, you touch on the aspect of *aiki* described in the previous section about Aikido. If you are capable of connecting to the center of another and feeling their rhythm, direction and energy, you can steer that person with very little effort. If a manager understands deeply what motivates each individual team member, he or she will be capable of leading the team with much less effort. If people have objectives that conflict with yours, understanding the core drivers of their objections is the best starting point for efficiently arriving at a joint effort or at least reducing unnecessary friction. Truly connecting to people is a powerful source of efficiently joining forces. You can only connect with people if you truly respect who they are and what drives them. Center control is the principle of alignment with your environment initiated from the heart of your strength.

Distance control

Distance control is about creating clear boundaries that others do not cross. Again though, this is a matter of balance. If a fighter puts the opponent at a distance that is a hundred percent safe, that means the fighters can't touch each other unless there is a big difference in height between the two. In this last case, the tall fighter will try to maintain the distance, while the short fighter will try to reduce it. Defining the right distance is a matter of taking the right amount of risk by increasing and decreasing the distance to your advantage at the right moments.

In business, distance control is about knowing what you can deliver and then managing expectations accordingly. If you raise expectations that you can't meet, it will result in too high a workload or in not being able to deliver. In the first scenario your wellbeing or other work you do may be affected. In the second scenario your client or end user will most likely not be happy, which will require extra attention from you afterwards.

The optimal distance is related to your speed of reaction. If a fighter sees an attack coming relatively late there is not much he can do apart from trying to block it (or even absorb it) and then recover by creating distance as soon as possible. Greater awareness allows a fighter to see an attack coming earlier, so he might be able to both block it and make an immediate counter attack.

The ultimate form of awareness enables a fighter to close the distance and attack right in the middle of the opponent's attack preparation. The opponent will be at the early stage of closing the distance to open an

attack and will therefore walk straight into the counter attack. This state of body and mind where awareness, balance, center control and distance control work in perfect harmony is called *Zen No Zen*. You basically counter an attack before it has the opportunity to reach its point of impact.

The earlier you see an attack coming, the more options you have for responding to it effectively. This is also true for professionals. If a client raises objections and you do not deal with the problem until he or she is really unhappy with your service, it will be harder to reverse the issue. If consumers are unhappy with your product, but find it easy to connect with your company and the response they receive is beyond their expectations, you might be able to turn the situation into an advantage. You could make these consumers so happy that they tell others about the positive experience. If it was hard to connect with your company, or if the service was lousy, it would take more time to repair the damage from these consumers talking down your product and company.

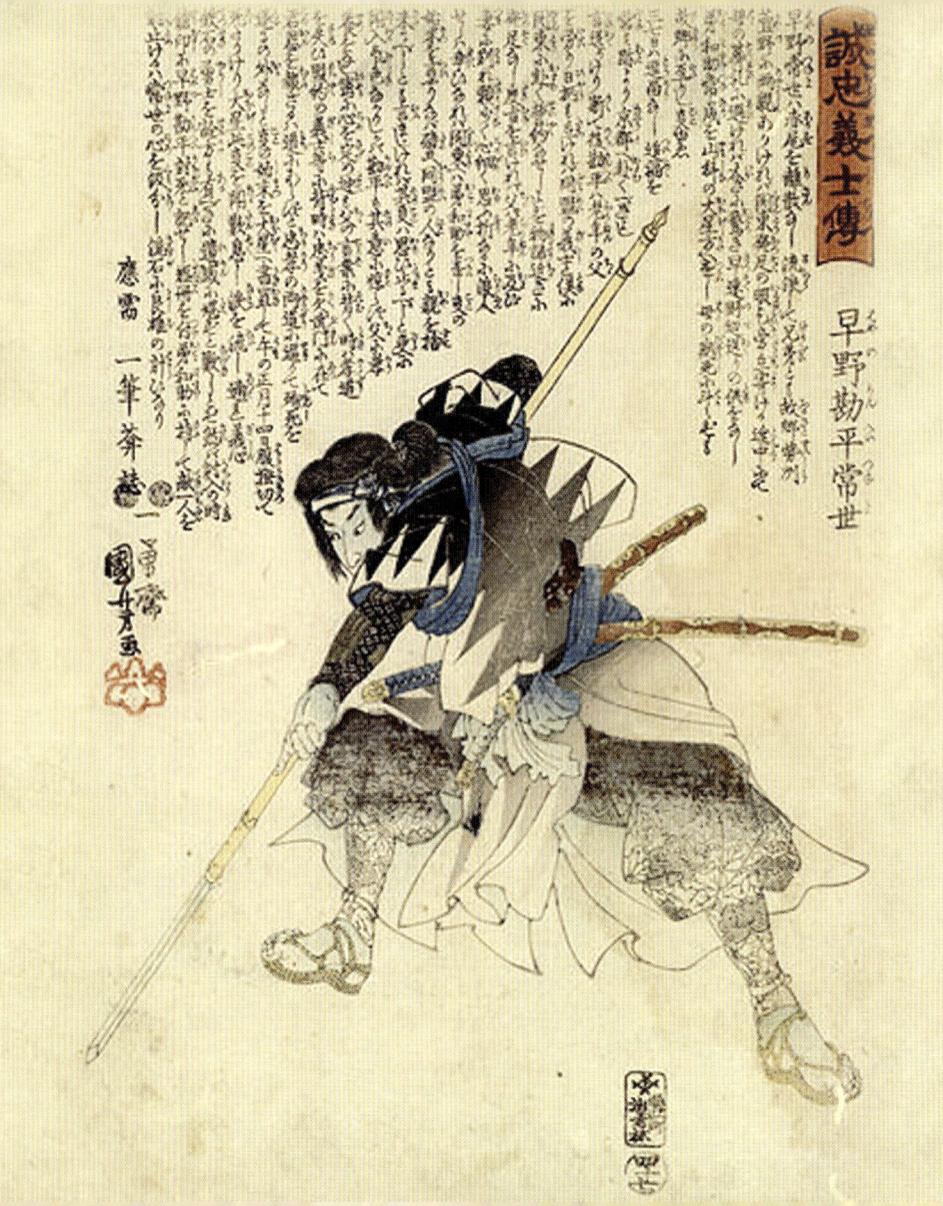

The principle of totality

The difficulty about managing the basic principles of fighting is you can't overlook any of them at any moment. If one of the four principles is not managed well, *all* of them will be affected. You can only reach your full potential as a fighter, as a professional or as a business, if you work from what I call *totality*. You need to be a hundred percent consistent in everything you do. If you are sometimes aware and sometimes not, it basically means you are *not* aware. If you are not aware, you cannot monitor whether your balance, center control and distance control are aligned with your changing environment. If your balance breaks, it diminishes your ability to establish center control and distance control. If your center control breaks, it diminishes your ability to establish balance and distance control. If you don't control distance you won't be able to manage balance and center control.

All elements need to be in place. Disturbances in any of the four factors need to be constantly spotted and neutralized. The more time you take to spot a disturbance, the more effort will be needed to neutralize it. The more time that is spent in neutralizing disturbances, the less impact your actions will have.

Let me explain how this principle of totality applies to business. If you are working with colleagues on a project and make an agreement to complete a list of actions, several things are important. First, everyone needs to be aligned on *why* the overarching project is important and what its objectives are (center control). Second, each person needs to be aware of how many actions he or she can reasonably take on (balance) and expectations should be managed accordingly (distance control).

In many cases a mixture of the above dimensions will not be aligned perfectly for all individuals, especially if you work in a cross functional team with different objectives. Some individuals might not agree that the project you are working on is important enough to spend time on. Maybe they think the project should head in a different direction, which is something they may not explicitly mention in meetings. Others might take on too much work and therefore not deliver on time. Others might have the objective of making sure *you* do all the heavy lifting and therefore put you under pressure to take on most of the workload.

Now, if you have not explicitly asked for agreement on the overall objectives of the project, you will not be able to confront people later if they don't act in line with them. If you don't deliver what you promised to deliver yourself, it will be harder to expect others to do so. If you don't manage expectations in the right way, you might end up having to deliver something you can't deliver. This means you will be overloaded with work, which reduces the time you can spend on making sure others deliver what *they* promised.

Everything is related and the principles apply on many levels of execution. You can expect people to read and respond to your e-mails only if you don't have a reputation for writing long or irrelevant e-mails or for not answering e-mails yourself. You can expect people to listen to your considerations, only if you don't misuse meetings for self-promotion. If someone agrees a deadline with you they need to know they will be chased if they do not deliver. If you are inconsistent in holding people to agreements, they will always know they can get away with not delivering. Finally, it is only worth chasing the agreements you make if you only make agreements that are important enough to chase.

You can't be sloppy in any aspect of your work. If there are areas where you feel you can be sloppy then reconsider whether they deserve your attention at all. If something is important enough to deserve your attention, it is important enough to have your *full* commitment. If you are a hundred percent consistent in establishing this system of discipline for yourself, you can gradually start demanding it from the people around you. The more consistent you are, the more efficient you can be in making sure people contribute to mutual projects. In the long-term most people working with you will feel happier if you work with a high level of consistency. Those who prefer not to commit will be less happy, but they will gradually disappear from the projects you work on and will no longer slow you down.

Coping with complexity and risk

Managing the principles of awareness, balance, center control and distance control with totality is hard, particularly because both fighters and professionals work in continuously changing and complex environments. Tracking all the movements of two bodies in a fight and assessing all the implications of those movements real-time, is a highly complex task. Tracking all your actions, those of your stakeholders, competitors and environment in business is equally complex.

It gets even harder when there is the pressure of risk. Imagine you are a fighter. You have practiced hard and in training you have built strong skills in sparring with opponents, assessing their movements and responding to them effectively and flexibly. Now you need to step into a cage to fight the ruling world cage-fighting champion or even worse, you need to step into a life or death battle. This pressure is likely to affect your ability to stick to the principles of fighting. Some fighters become stronger under pressure, while others are not even a shadow of what they were in training. Pressure can affect any of the four principles of fighting and mostly affects all of them at the same time, since they are interconnected. Some people start focusing on their anxiety, which means they overemphasize internal awareness at the cost of external awareness. Anxiety can create body stiffness, undermining balance and center control or the ability to move freely. Others try to calm themselves down by focusing on distractions, which might harm internal awareness or the ability to focus on *relevant* aspects of the environment.

Even though business is mostly not a matter of life and death and you generally do not risk physical damage, professionals are definitely under pressure in many ways. In many professions large amounts of money are involved or mistakes could have legal consequences. The potential loss of a job or promotion can cause serious feelings of pressure and stress. This pressure can make some people disconnect from their environment and make them aggressively chase their goals, without acknowledgement of their environment. In the best-case scenario this means you just don't work as efficiently as you could. In the worst-case scenario you harm the feelings or objectives of important people and therefore lose their support. Some people disconnect from *themselves* under pressure and keep chasing what they think are the expectations of others. Meanwhile they fail to notice their body is not responding well to the stress levels, or that their partner and kids at home might not be getting the attention they deserve.

Train yourself to stay internally and externally aware under any pressure. You can achieve this state of mind by balancing your challenges with moments of reflection and by taking good care of your mental and physical health (read more about this in the chapters *Challenge yourself, be persistent and reflect* and *Take care of yourself*). The ability to always be internally and externally aware, will allow you to keep your objectives and actions aligned with your environment (center control) and manage expectations (distance control) while remaining healthy and relaxed (balance).

Five rules for self-development

I mentioned in previous chapters that professionals should put self-development above the demand for success and money. This means that self-development deserves structural attention. In this chapter I offer five guidelines or rules that I find helpful in focusing on continuous improvement.

First rule: Be responsible
When fighters lose in competition or in training, they sometimes feel the urge to say "it was the referee's fault" or "my opponent wasn't fighting fair" or "he was too strong." In fact, all of these remarks may be true, but they are placing the blame for the loss on someone or something outside themselves. And external causes are harder to influence.

No matter what line of work you are in, if you want to develop your skills and personality, you need to realize that you are always the one creating the opportunities for good and bad things to happen to you. Instead of blaming someone or something else, it's better to look at how you can adapt your training or development strategy so you can better cope with problems in the future and become more successful over time.

Not everyone has the talent to become a world champion or CEO of a big multi-national, but it's not about becoming a world champion or getting the highest possible title, it's about developing yourself to the maximum of your potential and thereby being victorious over yourself. Everyone has unique talents and you will only discover your potential,

if you are always willing to reflect on what you can do better next time.

Second rule: Be proactive
When you live according to the concept of being responsible, you will always look at what you can do to make things better when they go wrong; being proactive means that you should do this even when things are just fine. Even if you win a match or get big promotions, you should actively review your performance, looking to see whether there are things that you can do better. Don't wait until you meet an opponent or competitor who exposes a weakness in your strategy – find them yourself. Always look for ways to improve regardless of whether you are winning or losing.

I will illustrate the principle of being proactive with the example of my fight with Gilbert Yvel in a Brazilian Jiu-jitsu competition. Gilbert Yvel is a strong Free Fighter. He won the Rings Free Fighting world title two times. At that time Gilbert was working hard on his skills in ground fighting. He joined competition in Submission Grappling and Brazilian Jiu-jitsu to gain experience.

In the hours before the match I thought hard about my advantages and disadvantages in relation to Gilbert. Gilbert was stronger than me; he was probably about 15 kilos (33 pounds) heavier and in better shape because he was a professional fighter. He was handy in groundwork and I knew he would be hard to submit. On the other hand we were fighting in *Gi* (Judo outfit), which was my advantage because I had practiced Judo since I was six years old. Also, I knew I was stronger in a standing position as long as I kept him at a distance. Letting Gilbert get in closer would enable him to grab me around my body, which is something he knew about from his Free Fighting experience.

With this knowledge I defined my strategy. I would grab his sleeve and collar at some distance and keep him there until I saw a perfect opportunity for throwing. From the throw, I would keep the advantage on the ground and only go for Submission (forcing to surrender) if I could make him tired. Submitting too soon would give him a chance to use his strength to take away my control of the position. This would allow Gilbert to use his skills in ground fighting and profit from his superior condition.

The fight went exactly as planned. I was very patient, keeping the distance between us and I managed to throw Gilbert and get into a favorable position on the ground. In this position I gave him just a little room to make sure he kept fighting and making himself tired. His tiredness came, but only shortly before the end of the game. I was ahead in points and decided not to take any risks. I kept the controlling position until the time was over. It was not spectacular, but it was tactically the right thing to do to win the match.

I profited from my advantages and managed to block Gilbert's. I was never in a position where he could profit from his strength, weight, condition or Submission Grappling skills. This is why this example gives a good insight into being proactive. If we had to do the same fight again the outcome might be entirely different because Gilbert would have learned about my strategy from the previous fight. Again, I would need to reflect on and enhance my strategy proactively.

If you want to work on certain projects or acquire certain skills, don't wait for people to come and help you. Find them yourself. Find the expert on a topic and discover what that person is working on. See if you can get involved. In many companies you only get promotions if you are already performing at the next level. You tend to only get the better projects if you are already halfway involved before you are formally selected. Sometimes it is even possible to create your own future job role by starting projects that are currently not supported by others or by finding new and better ways of doing things. This of course only works, if you can argue that your projects are valuable to the company. Being proactive helps you to create opportunities that otherwise would not have been there and it makes things run smoothly, where otherwise you might have bumped into a mess of problems.

The Burden

Two monks were returning to their monastery one evening. It had been raining and there were puddles of water at the roadsides. In one place a beautiful young woman was standing, unable to cross because of a large puddle of water. The older of the two monks went up to her, lifted her and carried her to the other side of the road. He then left her and continued on his way to the monastery.

In the evening the younger monk came to the older monk and said: "Sir, as monks, we cannot touch a woman?"

The older monk answered: "Yes, brother."

Then the younger monk asked again: "But Sir, how is that you carried that woman over the road?"

The older monk smiled and replied: "I left her by the side of the road, but you are still carrying her."

Third rule: Reflect without judging

Always trying to improve means being conscious about all your actions, about *how* you perform them and *why* you do it that way. You also must be conscious about your emotions, because if you are unaware of your emotions, they will affect your actions.

If you are critical of everything you do, you might get angry with yourself because you keep finding mistakes. If this happens, try to be conscious of this emotion. See why you are angry with yourself. Learn to see where you can improve without judging yourself for not being perfect. If you cannot do this you will be frustrated every time things don't go the way you want them to go and you won't be able to enjoy the things you do. Moreover, all the time you spend being angry with yourself is time not spent on improving yourself. This means you are wasting time and energy while increasing your dissatisfaction.

In Judo you often move through four phases of a learning process:

1. You enter competition, you get thrown and you have no idea what happened.
2. You see how the opponent throws you as it happens, but you still can't block it because you don't fully understand what happened in the preparation *before* the throw.
3. You see the preparations and you block the throw, but you still can't score or lead the fight.
4. You block and counter the attack successfully.

Learning in Judo is literally a matter of falling down and getting up many times over, and it is the same in business. The more you judge yourself when you fall, the longer it will take for you to understand what is happening and the more slowly you will move through the required phases of learning. In the previous story of the two monks and the woman, the young monk is preoccupied with a feeling of guilt due to the practical event of assisting a woman. This event harms his view of perfect obedience to the rules. He would develop further by reducing his judgmental nature.

How much you can handle varies from moment to moment and there are limits to what one person can do. Sometimes things go wrong and there is nothing you can do about it. When you see things going wrong, make sure you flag issues in the earliest possible stage and in a way that makes you confident the right people are aware of what is happening. In any company things will go wrong from time to time. Just make sure it's not because of unnecessary sloppiness or lack of learning from previous mistakes. If you can work like this, don't judge yourself for any remaining imperfections.

Fourth rule: Make your long-term goals explicit
Being aware includes knowing what you want to accomplish in your job, your career and your life. Many people have a hidden feeling that they want to be a champion in something or that they want to make a difference. If this is so, make it explicit.

Only if you make your goals explicit will you find the most efficient way of achieving them. Training to be a champion in sports is totally different from, for example, training because you want to be fit. If you aspire to a certain CEO role, you might need to consider moving abroad to increase your opportunities for success. Think hard about the investment it will take. Then decide if you still want to stick to the objective. You might discover that certain objectives require sacrifices you are not willing to make.

It can be hard to make certain choices when you have young children or are thinking of starting a family. Some people end up feeling overloaded with demands at this point, and may feel that they are not fulfilling all their roles adequately. It is perfectly okay to consciously decide to put your career into a slower lane for a while so you can focus on your children. It may also suit some couples for one of the two parents to maintain their speed in career development, while the other slows down. Just make sure this is an explicit decision based on mutual agreement. If decisions on long-term goals are not made consciously someone may suffer from it and be frustrated. It may be yourself having too much on your shoulders or it might be your employer or family not being happy with your efforts.

Fifth rule: Set priorities based on your long-term goals
Explicit long-term goals help you to set priorities in the choices you need to make on a daily basis. If you want to be a sports champion, you have to make a lot of choices. You need a club where you have enough opposition from training partners. You need a teacher and coach who

inspire you. You need to watch your nutrition, your training and your sleep. All this is only possible if you refrain from undertaking a lot of other activities. You should probably put school and work at a lower priority level, watch less TV, quit your second sport, drink only moderately, think about how much time you want to spend with your friends and your partner. Business objectives have the same sorts of trade-offs. If you want to undertake further education alongside your job you can expect to be away multiple evenings or weekends over a long stretch of time. If a business contact than asks you to dinner, that constitutes another evening away from home. Without the objective of further education, you might have said yes to the invitation. The objective of pursuing your education is already compromising your work-life balance, so the dinner might just be too much. Explicit long-terms goals will help you assess the value of each trade-off you have to make. This helps you to prioritize the choices that are fired at you throughout the day. Time is limited, so you can't be a champion in ten things at the same time.

Musashi writes: "do nothing which is of no use." You can best assess if something is useful, by looking at it in the light of your long-term goals. And don't forget: fun, inspiration and relaxation are useful priorities for almost any long-term goal. If I look back at my life as a Judo competitor with the benefit of hindsight, I am pretty sure I would have done better, if I had trained less and relaxed more. At that time I took the statement "do nothing which is of no use" a bit too narrowly, so I never stopped training. You should optimize the balance of your activities in the light of your long-term goals.

Iaido

Iaido is a modern Japanese martial art associated with the smooth, controlled movements of drawing a sword. The word approximately translates as *the way of mental presence and immediate reaction*. The primary emphasis in Iaido is on the psychological state of being present. The secondary emphasis is on drawing the sword and responding to a sudden attack as quickly as possible.

Many practitioners of Iaido use a real Samurai sword, with the edge blunted. Because Iaido is practiced with a weapon, it is almost entirely practiced using forms, or kata. Iaido does include competition in the form of kata, but does not use sparring of any kind. Because of this non-fighting aspect and Iaido's emphasis on precise, controlled, fluid motion, it is sometimes referred to as *moving Zen*.

If Iaido forms are performed alone, the practitioner fights one or more imaginary opponents. The more realistically a practitioner can envision the attacks, movements and surroundings of imaginary opponents, the more effectively the forms can be practiced and executed. If a practitioner does *not* have the ability to mentally visualize realistic fights in detail, the movements will look like an empty and fixed sequence of actions. With the right mental visualization the forms look like the practitioner is fighting real opponents, who make unexpected moves. So you could say Iaido is about the power of imagination and understanding of battle.

Plan your future with dedication and an open mind

So, it is important to know what you want to achieve in the long-term and it is important to prioritize your actions based on that. Success requires planning. Planning, however, has a downside: it assumes you can predict the things that will cross your path. If you talk to older people and hear about their career paths, you will often notice that many of the steps they took would have been impossible to predict at the start of their career. At many points they were offered unexpected opportunities. Something in them at each of these points decided that certain opportunities would be a good choice and then each choice led to new surprising future opportunities. In my specific case, if my knee had not been permanently injured, I might not be a research manager for Google now. Without the injury, every decision made in my life, would have been based on a different weighting of options.

So planning is important, but needs to be balanced with something that allows for flexibility. This is exactly the same in fighting. Preparing your fights is an important part of being a successful fighter, however, at some point in his career every fighter falls into the trap of planning a fight with the wrong level of detail.

I unconsciously experienced this problem for the first time when I was around seven years old. I did my first competition in Judo and of course was very nervous. In my eyes all the other kids looked stronger than me. I had a white belt with a yellow stripe while some of them already had a brown stripe or even a yellow belt. The referees looked big and serious and they were shouting loud Japanese words I did not understand. I walked onto the mat for my first match and a miracle happened: within the first seconds of the fight, I walked over to my opponent, grabbed him and was able to throw him on his back. I wanted to go on, but the referee touched my shoulder saying the match was already over and that I had won.

I was thrilled and tried to recall how I had made the throw. I wanted to do it exactly the same in the next fight. In the next fight, I immediately walked to my opponent and initiated the same attack that had been successful before. This time however, my opponent counter attacked. He scored a small point and the fight continued. I was puzzled. Something had gone wrong. I tried the same attack multiple times, but the more I tried, the more I failed. Of course I failed, because my opponent could see the attack coming from miles away, especially after trying it multiple times.

The truth is: every fight is different. Even another fight with the same opponent is different from the previous one. There are too many complex and unpredictable elements in any fight to be able to effectively plan in detail how you will win the match. If that could be done, the match would already be determined from the start.

So what should a fighter prepare for? Preparation should focus on defining the *battle area*, not on the action itself. For example, if you know your opponent is very strong and prefers certain grips, you can plan to block these grips wherever possible and do only small attacks until the point when your opponent starts getting tired. When that happens, you will need to assess what openings that brings you and find creative solutions for that. You might have suspicions about what techniques might work. You should however not focus on them so hard, that you start applying them when the opportunity is not really there, while at the same time other opportunities may pass without you noticing them.

I hope you have already found some parallels with the business situation, while reading the previous paragraph. You can plan the rough directions you want to take. The planning should be specific enough to rule out some options, so you can work more efficiently. Planning should however never be so tight, that you are no longer capable of reacting to unexpected problems or of grabbing unexpected opportunities.

As a manager you should give your team clear guidelines on what objectives to shoot for in the next period. You might even formulate some dos and don'ts in getting there. You should not interfere with the details of execution from the start. It takes away the feeling of ownership from your team members and also the ability to come up with creative solutions for unexpected problems or opportunities. Your employees might have insights based on their information or experience that you wouldn't think of from your management position.

If companies or managers fail to give clear strategic guidelines or keep changing them over time, you will see this affect the course of action in meetings during execution. You will notice that the same discussions keep recurring, without systematic progress. In a weakly prepared fight, this will be the fighter having doubts popping into his head blocking him from making fast and effective decisions. In successful businesses each new decision builds on the knowledge and results of previous decisions. Meetings have clear agendas and decision makers. Each action that is noted will be chased until it is done or until new information makes the action irrelevant. There can still be disagreement in meetings, however the meeting will always end with a decision, to which all members commit as if it was their own. Mistakes may still be made but since the right decision makers were involved there will be no useless finger pointing. The mistake will be evaluated and its lessons will be taken into account for future strategy and actions.

Complex topics or projects always require iterations based on the latest knowledge. At certain stages of a project it is common to worry about whether the strategy is still on track. Particularly in complex multi-year projects, it is impossible to overview all the actions needed to make the project a success. It is important though, that worrying does not affect your ability to operate effectively. If you doubt strategic choices too often you will be delayed by every obstacle you encounter. The solution is staying goal oriented. Choose overall objectives wisely and commit to them. If you do, you will notice more quickly when your thinking isn't adding to solutions, but instead has crossed the line towards useless worrying. With each step forward, you will see new opportunities opening. There is a saying: "everything becomes fluid under pressure." If you work on the right objectives, commit to finding creative solutions and things get urgent enough, there will always be a way to succeed. In a fight, it can sometimes feel like there are no openings to attack at all. It may take many small attacks and a lot of patience, until one small opening reveals itself. If it does, make sure you are ready to take it. If the opening does *not* come, at least you know you gave whatever you had. You can then accept more easily that your objective may not have been feasible. Regardless of success or failure, your trials and experiments will give you a lot of food for thought in evaluating and improving your strategy.

Be aware that planning is always a trade-off against flexible execution. Plan sharply enough to allow for prioritization, but flexibly enough for there to be room to find creative solutions when problems or opportunities arise. As in the art of Iaido, you need to balance mental preparation with action. Don't plan an exact career path, but plan multiple desirable scenarios and then weigh the options that are presented to you against these scenarios. Step back once in a while and evaluate whether you are still performing effectively. Check whether your view of the desirability of each scenario has changed over time. Once you have done this, trust in your decision and commit to doing what is necessary to get where you want to be.

Managing your job happiness

Though it is important for companies to invest in keeping their employees happy, I also consider the management of personal job satisfaction to be the responsibility of any professional. It is your responsibility to let your manager and colleagues understand what you want. You need to flag problems or things you don't like as early as possible. If you don't do this, you deprive your manager and the company of the opportunity of helping you and keeping you happy.

You may not always immediately get what you want though. If you don't get something that is absolutely critical for your job happiness and you are a hundred percent confident you have been clear on an issue, make sure you flag it even more clearly. If necessary, add that this is crucial for you to want to continue working for the company. If you still do not get the changes you need to make you happier consider changing jobs, BUT be aware, no job is ever perfect.

A certain level of acceptance of things you don't like is important in every job. In the early stages of your career it can sometimes be hard to know how much you can demand and how much you should accept. If you continue feeling doubtful changing jobs might be the best solution. Two things can then happen:
1. You discover that things are indeed better somewhere else.
2. You discover some problems are present everywhere or different places just have different problems. Knowing this more deeply from experience makes it easier to accept unavoidable things you don't like.

From this perspective, a job switch can be useful, even if annoying things from your previous job are still present in the current one.

It is important to understand some contradictions here. On the one hand, the best way to be successful is to do something you like. At the same time, any job will inevitably have some elements attached to it that you *don't like*. Examining each element of your job to determine whether you like it or not will waste time and energy, making you slower and less successful. You should accept any inevitable unpleasant parts of your job as just being a part of it and operate with maximum accuracy and dedication. You can only do that if, on a strategic level, you think hard about the importance of your current job for your long-term goals. If you are convinced that your current job is the right one, dedicate yourself to any tasks that belong to it. If you doubt that the current job is the right one, flag it with your manager or start looking around for another job. If you have decided you need to look for another job and there isn't one available at short notice, keep doing your current job with dedication. That maximizes opportunities for your current employer to surprise you with an opportunity, particularly if you are open and clear about your job expectations and career plan. In most cases, I have been open to my employers in moments where I felt my job wasn't challenging anymore. It gives an employer the opportunity of facilitating your objectives. If they don't support you in such situations, you can be sure that you need to look for another job.

Another contradiction can be found in aspiring to a higher salary or promotion. I stated earlier that I firmly believe professionals should learn to value self-development over success and money. If they do, they will produce better output. Once your output is better, in an ideal scenario, your employer will reward you with a salary increase and career development. Unfortunately not many employers offer this ideal scenario.

In many cases you may not automatically get what you deserve because it is hard for your superiors to assess all the things you are doing. In any job, every now and then, you need to take some time to inform your manager of all the great things you are doing. If you go over the top this becomes annoying and useless self promotion (which is, unfortunately, still successful with some managers). However, there is nothing wrong in displaying your successes with pride. It helps your manager and the people around you to stay up to speed with your work and the skills you are developing. This will enable them to better assess what future projects or job roles you are suitable for.

Sometimes, you don't get the rewards you deserve because your employer has an opportunistic view of working with employees. In this case, as long as an employee doesn't demand a higher salary by threatening to go away, he or she is considered to be a cheap worker. You may wonder if you want to work for an employer like that. There might be good reasons though to stay with the company, for instance, because the atmosphere with your colleagues is good or because there still is a lot to learn. You will need to push hard in this case to get the pay raise and promotion you deserve.

So the ideal scenario is one where your focus on self-development is

automatically rewarded, however, most scenarios are not ideal. This means you need to be assertive in getting what you want. My teacher Jan often mentioned three guidelines for coping with things you don't like. I still find them useful today:

1. **Change it:** this can mean convincing others to see things differently or finding a way to work around it so that the problem is no longer there.
2. **Accept it:** accept that the problem cannot be solved and try not to be bothered by it.
3. **Leave it:** walk away from the situation, which in this case would mean you look for a job in another company or in another team within your company.

If you don't choose between these three options, you will find yourself moaning every time the same issue arises, which will make you less effective and will be annoying for the people living and working with you.

Take responsibility for your job happiness. You spend many hours of your life in your job, so don't settle for something that makes you feel unhappy, unless it is unavoidable, in which case you may just have to accept it, at least in the short-term.

Jiu-jitsu

Jiu-jitsu was developed around 1532 by Hisamori Takeuchi, a military tactician. Hisamori combined various Japanese martial arts that were used on the battlefield for close combat in situations where weapons were ineffective. Jiu-jitsu has gone through different phases, changing as strategies of warfare have changed, forcing fighters to utilize different techniques or weapons. As a result Jiu-jitsu has changed significantly through the years. Because striking against an armored opponent proved ineffective, practitioners initially focused on techniques like pins, joint locks and throws. These techniques were developed around the principle of using an attacker's energy against him, rather than opposing it.

In the early seventeenth century, strict laws prohibited various types of armor with the aim of reducing war. During this period, weapons and armor became unused decorative items and hand to hand self-defense flourished. New techniques, including striking techniques, were created to fight unarmored opponents. However towards the eighteenth century the number of striking techniques was severely reduced as they exerted too much energy. Instead, striking primarily became used as a way to distract opponents or to unbalance them in preparation for a joint-lock, strangle or throw. In contemporary Jiu-jitsu we can find the remnants of all phases of development. There are classical kata's with punches, kicks, throws and weapons. Some forms even refer back to the weak points of the armor of an ancient Japanese soldier. It is intriguing how a martial art can evolve due to the need for soldiers to survive in combat.

Don't fight inevitable developments

In the history of Jiu-jitsu we see how warriors evolved in a changing environment due to the need to survive in combat. It was often said that the true indicator of a Samurai's skill was his age. Adaptive skills are equally important for professionals and companies. Big shifts like the development of the Internet and computers are still changing the dynamics of business and therefore the required skill-set of professionals. Almost everyone's job has been affected by these two phenomena. Nowadays, if you don't know how to operate a computer there are very few jobs you will be able to do.

As techniques evolve, certain job roles become redundant and others emerge. In the same way, the relevance of products or services can diminish, increase, appear or disappear. Companies and professionals should therefore always monitor whether the inevitable developments are likely to affect jobs or business in any way. Very often companies choose to fight developments, as we saw in the example of the music industry fighting downloads.

Many traditional companies say the Internet messes up their business. It depends on how you look at it. If we take the book market as an example, it is true that many bookstores have shut down because consumers are buying books online and because some people now only buy digital versions of books. You could argue that this damages the culture of cities as bookstores are pleasant places. It's nice to walk into stores, browse through books and be inspired. On the other hand, thanks to the Internet anyone can publish a book, since shelf space is no longer a limiting factor. Variety in books is greater than ever and book prices have gone down, so knowledge and culture have become richer and more accessible to more people. Bookstores can still be part of the business model, but they need to be combined with online stores. I will not be surprised if, at some point in time, companies like Amazon start opening offline bookstores as places of inspiration. Those stores would then have a similar role to that of the flagship stores of Apple, bringing the product experience closer to people. The rules of the game have changed, which means existing strategies need to be adapted or replaced by new ones.

If a development is inevitable, time spent fighting it is time not spent in adapting. There is no escaping this in the long run. Almost every master that has established a new martial art stresses the importance of accepting and redirecting a powerful force, rather than opposing it. This skill is a must-have ingredient for efficiently getting to where you want to be. Every once in a while, some fundamental change affects your business or job and you need to make sure that you capture its implications as fast as possible when this happens. There is no choice on these occasions, it is just a matter of timing. If you adapt too early, you might end up killing a cash cow unnecessarily or end up investing in a development that is not becoming big after all. If you adapt too late, you potentially risk losing your whole business.

In relation to your career, it is important to monitor whether developments will require you to build new skills over time. If you react to changes too early you may end up investing time in skills that may not be needed after all. However, if you react too late you may end up losing your job because your skills are no longer relevant. I personally feel that the need to keep developing new skills is what keeps a job fresh and energizing over time.

Managing chaos effectively

Many professionals need to juggle multiple projects or actions at the same time. In my own profession I always have at least fifteen research studies running at the same time. These projects are generally at different stages, and most have multiple stakeholders requiring different approaches in terms of tactics and execution. Someone like a receptionist, for instance, also regularly needs to juggle many projects. A receptionist may receive multiple calls at the same moment that the postman arrives with a package and the director wants to reschedule a twenty person meeting. People who run their own businesses also need to multi-task. Besides their core business, they need to think about finance and taxes, about maintaining their website and things as simple as answering the phone. They often find themselves having multiple jobs at the same time.

Managing a lot of projects simultaneously has many similarities to fighting multiple opponents at the same time. When fighting multiple opponents your first priority is to be proactive. If you wait for your opponents to come to you they will all come at the same time. *You* should be the one who chooses which opponent to take on first. Then each time you handle an individual opponent, you need to make sure you handle him so effectively that he is not likely to stand up and attack again while you are busy with your next opponent. During your encounter with one opponent, you must make sure the rest of the group stays at a distance by choosing your position wisely. All these actions refer back to the basic principles of fighting effectively: awareness, balance, center control and distance control.

If you translate this to being a professional, you will find the parallels. If you are too passive as a professional, people will approach you all at the same time, with all kinds of tasks that are relevant to *them*. You should define very clearly what your objectives are, what tasks *you* want to work on and then align those with the objectives of the people around you (center control). Once this is done, you should manage expectations (distance control). Be clear in *what* people can expect from you and *when* they can expect it. You need to stay open for new requests coming in, but always align against *your* objectives (internal and external awareness). Based on this alignment you can prioritize and manage expectations. While you perform all these tasks, you need to continuously monitor whether you still feel comfortable handling the things that you promised (balance). If someone with a sudden urgent request wants to take time from you, blocking you from delivering other promised actions, don't dive in like a madman. Consider delaying other things first (managing expectations) or asking for help and then attend to the matter. If the matter is so urgent and important that it can't wait for a few minutes, you should immediately attend to it and manage others' expectations afterwards.

There are basically two types of tasks: tasks that require your full attention now (the opponent you are fighting) and tasks that do *not* require your full attention now (the rest of your opponents). For tasks that do *not* require your attention now, you should pay just enough attention to assess their priority and weight: *when* is your attention needed and *how much* effort will it take? Sometimes you also need to assess if others should make your tasks their priority: *who* do you need to involve? Once those aspects are assessed accurately, you can leave the matter. Any task that requires your attention now, should have your *full* attention and should be handled thoroughly. If you fail to be thorough, the task will come back later and will then mess up your planning control. If a task is not important enough to require your full attention consider whether it deserves your attention at all.

If you can handle tasks like this, with discipline and dedication and if you take some moments for yourself from time to time to mentally and physically relax, you will be surprised by how many projects you can handle without becoming stressed.

Coping with tactical resistance

When groups of people work together and career success or money are involved, political behavior will arise at some point. It is in the nature of people to fight for their personal objectives and not everyone fights as fairly as you would wish. As I mentioned before, tactical behavior can come from both outside your company (competitors for example) or inside your company (colleagues or other teams competing for the same resources and rewards). Sometimes you expect political resistance and sometimes it can take you by surprise. Either way, you should be ready for it.

I discovered one rule in coping with tactical resistance by doing hundreds of presentations explaining the implications of research projects to marketers. As a researcher, I was being paid to evaluate the performance of the advertising and marketing strategies of big advertisers. We would for instance check whether a TV commercial had been effective by measuring whether people noticed the ad, understood it and linked it to the advertised brand. Marketers are not always happy to hear that they have spent a few hundred thousand dollars on a campaign that wasn't effective. To me this always seemed strange, for I knew that most campaigns are not successful right away and that some iterations are usually needed to make them more effective. However, the superiors of many marketers don't understand this and might therefore have very low tolerance for 'mistakes'. Some marketers could even lose their jobs over my research results.

When people don't like research outcomes the standard response is to attack the numbers. Of course an attack can't be done openly, since that would expose the fact that the marketer is trying to hide the negative outcomes. The attacks are therefore always wrapped in the form of asking for methodological details with the aim of finding a weakness in the data that would justify throwing away the results altogether. In my earlier years, I always thought that people were sincerely interested in the methodological details and therefore nicely took the time to clearly explain everything. Luckily I always look at the same data from multiple angles before making conclusions, so I always succeeded in gaining acceptance of the outcomes. I noticed that in politically loaded situations, helping people to understand the data took a lot more time than usual. I discovered that this was because people weren't occupied in trying to understand things, but were occupied in looking for flaws. I was therefore connecting to them in the wrong part of their mental state.

The more I spoke to marketers, the more I understood the pressure they were sometimes under. I learned that I could come to agreement much faster if I addressed the tactical reason for a question, rather than the question itself. This is particularly important if you speak in conferences. Someone might want to misuse the platform to criticize issues that you cannot reasonably explain to a wide audience in the five minutes you have to take questions. If you then attempt to start your explanation without addressing the tactical nature of the query you will not be able to finish your story and will therefore leave the audience with a burning question. Politicians are very skilled in seeing through such tactics. They *must* be skilled at that because many journalists have a habit of creating news from the negative extremes of topics that have nuances that are hard to explain.

If someone asks you a question or makes a remark, you should not only listen to *what* is said, but also to *why* it is said. If the "why" is tactical in nature, you should consider not answering the question itself, but instead investigate the why. Try to understand why a person might have reason to object to what you are saying. You might be able to find a solution for that problem. In my example about marketing research, I discovered I could remove a lot of resistance by giving trainings to a wider audience in the company. In those trainings I explained that the majority of all campaigns are not effective in the first iteration. This means a strategy of launch and iterate is required to be systematically successful. You can only reiterate if you allow for failure, so you can't fire someone for having research that shows the campaign needs to be optimized. I would rather suggest firing someone who tries to *hide* negative results and does not learn from it.

In fighting, you also need to respond to an opponent tactically. In Judo for instance, there are fighters who deliberately do nothing in the first part of a fight. This means you will mostly not have openings for attack and at the same time will not be attacked yourself. The fight can become very boring after a minute or so and the audience will start reacting negatively. If you can't handle this pressure, you will start making risky attacks that can be easily countered. If your opponent is not engaging in the fight, you should not expose yourself to unnecessary risks. You should remain patient. Luckily, in Judo there are penalties for passivity. So an opponent using this strategy is likely to be punished. In life and death duels in the period of the Samurai there was no referee though. That could result in two opponents circling around each other, doing nothing for more than an hour. One of the two would get impatient first or his focus would snap and then suddenly, the fight could be over in only a few seconds and strikes.

A tactical win is sometimes more achievable than a spectacular knockout, particularly if you are in a business that is structurally going down (selling CDs for instance), it is important to know what you can realistically achieve. A friend of mine works for a bank that is facing tough times. I asked him how he copes with the idea that there are not really any successes to be celebrated in his company. He explained that his objective is to facilitate a soft landing for the company. I think this is very realistic. It is tactically the right objective.

In general I believe it is important that you have as little as possible to hide. If you are structurally open and honest in your objectives and how you approach them, it will be harder for tactical people to tackle you. Be sincere and cooperative, but very clear in what you want to achieve and what the necessary steps and time lines are. Consistently chase people to deliver what they have promised you. Align the objectives you work on closely with company goals. If you do that and you encounter tactical resistance, you are more likely to get senior help if you ask for it. I've said it before: if you work on the right things, everything gets fluid under pressure. If your superiors are not giving you support but you are confident you are working on projects that are important for the company you might want to rethink whether you are still working for the right team or company.

A lot of people suffer from the fact that routines and politics make them unhappy in their jobs, at least to some extent. If you can inspire other people they will enjoy working with you and will bring contributions to mutual projects. Don't be shy in sharing your knowledge. Give it to people and you will be rewarded. However, don't let yourself be lured into sharing all the information and knowledge you have if someone is being secretive towards you. Only spend time on people who will use the knowledge you share for the right objectives or for the benefit of the company.

Wearing different masks

In many ways, integrity is one of the most important attributes for a professional. But how does that match with having to play a diversity of roles? Depending on who you are talking to, you may be colleague, boss, client, competitor, mother, father, kid, friend or multiple roles at the same time. As your roles change from moment to moment, you may be required to communicate different messages to different people. How can you reconcile that requirement with the fact that you are still the same person aiming to maintain integrity? I believe integrity is the thing that keeps you rooted, independent of the roles you need to play. The more you are rooted in your personal integrity and identity, the better you can perform all your roles without losing yourself.

Personal integrity is the energy that drives true productive behavior in all situations. You can have an angry conversation with a poorly performing employee, even though you may be in a good mood. Being friendly because of your good mood would have given the wrong signal to this person, which would not have been beneficial for either of you. Your respect for the person in front of you is what will always drive you to give that person the message needed to clarify the situation for both of you. This can mean you give negative feedback about the performance of someone who normally performs very well, but made a mistake in a recent project. That person deserves honest feedback to maximize the opportunity to develop his or her skills. At the same time, one piece of negative feedback will not undo all the previous good work.

If people can always trust you to give honest feedback, they don't have to spend time wondering what you mean by what you are saying.

So your personal integrity needs to help you remain balanced while doing the things you need to do, while putting on friendly, angry, sympathetic or happy masks when required. This has similarities with the Samurai who wore masks when going to war. The masks were designed to protect the Samurai's face, but were also intended to make them look scary to their opponent. However, behind the mask was always the person who was quiet and alert, confident but not arrogant, trusting in his skills, but also aware of his own weaknesses – everything in exactly the right balance to ensure the best chance of success on the battlefield.

If you are more balanced, you have fewer limitations in performing your roles. You will have to pretend less and can be yourself more. A skilled fighter can employ his personal techniques in battle; techniques that are fully tailored to his body and skills. Less experienced fighters need to settle for copying techniques they have seen others perform. Find your own style in coping with the roles connected to your profession and you will be freer and less artificial.

Self-control

During the civil wars in feudal Japan, an invading army would quickly sweep into a town and take control. In one particular village, everyone fled just before the army arrived – everyone that is, except the Zen master.

Curious about this old fellow, the general went to the temple to see for himself what kind of man this master was. When he wasn't treated with the deference and submissiveness to which he was accustomed, the general burst into anger.

"You fool," he shouted as he reached for his sword, "don't you realize you are standing before a man who could run you through without blinking an eye!"

Despite the threat, the master seemed unmoved.

"Do you realize," the master replied calmly, "that you are standing before a man who can be run through without blinking an eye?"

With power comes responsibility

The monk in the previous story puts the angry general in an awkward position, by pointing out the clearly superior strength of the general. It would be cowardly for the general to attack the monk after that.

As you progress in your career, you will gain more status, power and tactical skills. The temptation to rely on those will therefore be larger. If your personal integrity is not strong enough, you can become addicted to that kind of power. Try to remain as selective as you can in exerting power or tactics, even when under pressure or short on time, otherwise they may start to work against you.

This is also the philosophy of martial arts. Martial arts are not a matter of aggression. The true art is in knowing that you can be aggressive and assertive if it is required. Knowing this deeply will make you more relaxed and capable of handling difficult situations in a way that is friendly, patient and sincere, while people will feel your underlying determination and commitment. The result is an increase in speed of handling projects, tasks and issues and at the same time a reduction of stress for both you and those around you. The knowledge of your capacity to succeed in potential physical or mental battle will result in having fewer conflicts. It will help in creating an environment of cooperation and respect around you. This respect does not need to be enforced. It is just there, because you are who you are and you do the things you do.

The best win is one in which the loser does not feel defeated. The best meeting is one in which you reach your goals and all the other people in the room have the same feeling. That means you need to allow others to reach their goals as well. This principle applies to business goals, personal goals and the emotional goals of the people you work with.

If you lose yourself in tactical behavior without inspiring and convincing people with sincere arguments, you will lose their support. Keep nurturing your will to help people in reaching their full potential. Even if you have developed skills to attract the better projects to work on, don't refrain from supporting the people who are doing the 'dirty' work.

Be as selective as you can be in putting on masks, otherwise people will no longer be able to see who you are. You will lose their respect as a result. Who you really are is always the source of your success. Masks may be needed to take you through some challenges every now and then, but don't put yourself in positions where you need to say things that conflict with your self. Make sure you don't harm your credibility as a human being. If you stay true to yourself and can refrain from exerting power or tactics when not needed, you will get maximum support from the people working with you.

剣
道

Kendo
"Way of the sword"

Kendo

Kendo means *way of the sword* and is a modern Japanese martial art based on traditional Samurai swordsmanship. From the time of the earliest Samurai government in Japan (1185–1333), sword fighting, horse riding and archery were the main martial pursuits of the military clans. In this period, Kendo developed under the influence of Zen Buddhism. A Samurai should be able to have disregard for his own life in order to succeed in the heat of battle. This notion was well-matched to the Buddhist concept of the illusory nature of the distinction between life and death.

The introduction of bamboo practice swords and armor to sword training is attributed to Naganuma Shirōzaemon Kunisato, who established his own training method. Kendo is still practiced with bamboo swords and armor. Kendo derives concepts from Zen Buddhism, such as *Fudōshin* (unmoving mind). *Fudōshin* implies that the Kendōka cannot be led astray by delusions of anger, doubt, fear or surprise, which are collectively called *the four Kendo sicknesses*.

Kendo training is quite noisy in comparison to some other martial arts or sports. This is because Kendōka use a shout (*Kiai*) and a stamp of the front foot (*Fumikomi-ashi*) to express their fighting spirit when striking. A strike or thrust must hit specified target areas and is only valid if the attacker demonstrates fighting spirit, correct posture and if the score is followed by *Zanshin*. Zanshin is the continuation of awareness: the Kendōka must be mentally and physically ready to attack again.

Facing your fears

We all have fears, but while some fears are functional and warn of danger, many actually slow you down or limit your development. For example, I wanted to teach myself tactical skills to manage my workload better. I discovered I was afraid to say "no" to people. When reflecting on this, I felt this was linked to two fears: the fear of not being liked and the fear of doing things wrong. These two fears had a desire attached to them as well: wanting to be special. If I was special, I was (temporarily) safe from having done things wrong. When I saw this, I could also see that people misused both fears to increase my workload. They would mention small things that made me feel that I was in their debt. At the same time, they would link their request to a compliment: "you are the only one who can do this." Being aware of my fears helped me in seeing through these tactics and in saying "no" when needed. I thereby learned to manage my workload better. Consequently, even though I did less, people showed more appreciation of my efforts. It was no longer a given that I would help them.

I also discovered I was afraid of raising my voice in front of an audience. This fear was easily confronted by the fact that I had to do about three presentations a week to groups of marketers. I discovered that people were actually very happy to hear my opinion and the more I spoke the more people asked for my opinions again. Nowadays I present at conferences regularly and with joy. I sometimes still find it hard to say "no" to people, but then again my helpfulness is highly appreciated amongst my colleagues and I feel I am being rewarded for it. I sometimes still come to the point where I feel I helped someone who opportunistically used my time for their own benefit, but then again, everyone deserves a fair shot and once you have given that shot and it's been misused you know what to expect from those people in the future. Dealing with fears is often a matter of accepting the idea of potential loss. There is a counter intuitive relationship between being willing to lose and the capacity to win. Knowing deep inside that you will fully accept loss if it is inevitably forced upon you will increase your endurance in trying to win. I will explain this in the context of fighting and then move to situations, where this principle also applies to business.

A fighter needs to remain calm and focused every millisecond of the fight, even in extremely unfavorable positions. Imagine yourself being a Samurai in a sword combat situation. You are facing an enemy you don't know. The fight has already lasted for some time and you are clearly well-matched, which means you don't know if you will win this fight. You also know that if you don't win it, you will most likely die. During the fight, your opponent surprises you with an attack and you leap back just in time. The point of your opponent's sword reaches your throat and leaves a shallow cut.

At this point, if you have any fear of losing the fight or of dying, your mind will start playing tricks on you. Just before being hit, the knowledge of being a bit late in defending the blow will make fear take temporary control over your mind, which means you are no longer focused.

Directly after that, you realize the attack was not fatal. That positive surprise contrasts with the fear giving you a feeling of relief, which again means you are not focused. These two moments of fear and relief swing your awareness towards two extremes, both taking you out of combat, which may mean your opponent's next blow takes your head off.

This sequence of events can also work the other way around, which is even trickier. During the first five minutes of the fight you may feel strong. There may be multiple moments in which the attacker is in difficult positions. If you have any fear of losing or dying, each time this happens, your mind starts shifting towards the end of the fight. In your first strong moment you might think consciously or unconsciously: "I might be able to win this fight." In the next strong moments, that small thought gradually builds up and takes over a growing part of your awareness. Then suddenly your opponent surprises you and makes that first harmless cut. Your visualization of winning the fight shatters in a split second, takes you off guard and the final chop may be the result.

In either of those situations, a hundred percent absence of fear would have allowed you to keep a clear focus throughout the whole fight. The first scratch would have been nothing more than a scratch and the fight would have gone further. Now, if your opponent really is stronger you may still lose the fight, but eliminating fear and keeping a clean focus, even in threatening situations, will take you to the highest achievable point given your skills at that time. The longer you can stay focused the more likely it is that you may suddenly find that one opening needed to finalize the fight to your advantage.

For most people, work is not a matter of life and death. However, normal fears like the fear of not being liked or of doing things wrong trigger the same mechanics in your mind that the Samurai faced in battle. You can face your fears by visualizing the worst-case scenario of any event and then trying to reconcile yourself with the implications of that scenario. There have been many moments where I could relax, thanks to the idea that I would be willing to quit my job, for instance, if I was unable to reach agreement with my employer on certain objectives or boundaries. Knowing exactly how far you are willing to go and accepting that worst-case scenario, makes you calmer and more confident and therefore makes it less likely that the scenario will actually play out.

One thing is very important: not fearing loss or death doesn't mean you don't care about life. It is easy to remain unafraid of dying if you feel suicidal anyway. It is easy to quit your job if you don't care about it. It is easy to let your company go bankrupt if you don't care about the business or the people in it. That is a different way of not fearing loss or death. Not caring just means you won't try hard. The Samurai in combat cares a lot, but still does not have the fear of losing or dying.

I believe the ultimate point of personal growth is being able to meet victory and loss in exactly the same manner: by being grateful for the experience you gained. Any meeting, debate or project you participate in is a win-win experience. If you can invest a hundred percent of your time and effort in becoming the best you can be and still cope with the outcome of events as I've just described, you have reached the highest level of personal growth possible. This insight refers back to the concept of Zanshin, as described in the previous section about Kendo. If you have an unmovable mind, no matter what happens, you will always mentally and physically be ready to attack again.

There is one sneaky but common variant of not facing your fear. Sometimes, focusing on work is exactly what helps you to escape your fears. For instance, after a relationship break up or death of a loved one some people start working harder. The work focus reduces the feeling of pain. That can be useful, but can also be a trap. If you find you have trouble slowing down in work, you may have fallen into that trap. I noticed at one point that I always found a reason to work late. When I decided to do that less often, I became restless. When I saw this, I knew that by working hard I was covering up some fear. I had been working hard for many years though, so the fear would not show itself without a fight. I committed to not working late anymore, until I knew exactly what was causing the restless feeling. I went home at six in the evening every day for three months, but then realized that I had started checking and replying to work mail at home when feeling restless. I committed to not checking my work mail anymore in the evening for three months.

I discovered a fear of not producing enough to be sufficiently noticed by managers and team members. In the years when I worked extremely hard I was clearly the most productive person in the team. People were cheering my results and I regularly won awards, all of which gave me a feeling of security. However, I became dependent on that continuous positive feedback, which is unhealthy. In the six months that I forced myself to work less I felt my fear of not being good enough grow. At the same time however, I could see my colleagues were still happy with me. I was not getting constant applause but I remained a respected team member and even became more a part of the team. Seeing that this was possible made me more relaxed and freer to choose when I wanted to work late and when not. In total it took me about a year before I trusted myself to not fall back into working too hard. After this year, I decided my peace of mind was strong enough to start writing this book.

There is nothing wrong with working hard and being successful but from time to time it is worth checking whether you have become addicted to it. Skills you have can always be applied more effectively if you can freely choose when to use them and when *not* to use them. In the sword fight I described earlier, the overdrive mode would be a fighter who shuts down his fears and then fights wildly to keep them quiet. That can be successful, but he will lack the ability to strategically assess when to up the pressure and when to release it. He will therefore lack the element of surprise in his fights and waste more energy. A smart opponent will see this overdrive mode and strategically use that knowledge, for instance by wearing the opponent out before counter attacking.

The dragon symbolizes your deepest fears. If you face them, everything you do will be more effective. You will lose less time on doubts and will handle bigger and more complex projects. Do not attempt to kill the dragon and don't run away. Both reactions will make it stronger. If you try to erase or avoid deep fears, they grow. Train yourself to look the dragon in the eyes, feel the fear, doubts or pain, while maintaining focus and doing what you think is right and necessary.

空手

Karate
"Empty hand"

Karate

Karate is a martial art developed in the Japanese Ryukyu Islands (now Okinawa). It is a striking art using punching, kicking, knee and elbow strikes and open-handed techniques such as knife-hands (the famous Karate chop with the side of the hand). Grappling, locks, restraints, throws and vital point strikes are taught in some styles. Karate is often translated as *empty hand* referring to the unarmed style of fighting.

Early forms of Karate already existed in the thirteenth century. A large group of Chinese families moved to Okinawa around 1392 for the purpose of cultural exchange. They shared their knowledge of Chinese arts and sciences, including martial arts. The ban on weapons enforced in Okinawa in 1609 further stimulated the development of Karate. In 1924 Keio University established the first university Karate club in Japan and by 1932 all major Japanese universities had Karate clubs.

Karate became popular worldwide due to movies and TV shows displaying Karate as a mysterious martial art capable of causing death or injury with a single blow. Shigeru Egami, Chief Instructor of Shotokan Dojo, stated that "the majority of followers of karate in overseas countries pursue karate only for its fighting techniques" and that "mass media present a pseudo art far from the real thing." Shoshin Nagamine (another famous master) described Karate as "a life-long marathon that can be won only through self-discipline, hard training and one's own creative efforts."

Challenge yourself, be persistent and reflect

True self-development often requires fundamental improvements or additions to your skill-set. Fundamentally new skills rarely come quickly or with little effort. People looking for the shortcut will eventually see their learning curve flatten out. Diligence and patience are always needed and you will be challenged in many ways along the road. Think about the prisoner's dilemma of office politics I mentioned earlier. You will see other people taking shortcuts and they might even be successful in the short-term. You will need to continually remind yourself that you are on the right track of self-development to reach your potential in the long run. Unless you are very lucky, you won't succeed without that kind of dedication, because you won't be able to stay in a continuous flow of self-development.

In addition to dedication, you need creativity, for there will always be unexpected barriers. Projects that seem easy before you start can turn out to be a lot more complex than expected. Force yourself to keep looking for creative solutions. If you work on the right objectives, you will find effective solutions.

Fundamentally new skills cannot be learned in theory only. They have to be acquired by doing, experimenting, evaluating and trying again. When practicing meditation, my teacher always said: "you have to bring meditation to the market place." It is not sufficient to be able to have a quiet mind while sitting on a beautiful mountain. True meditation is doing the things you need to do while being aware. This takes a lot of fluffiness out of the concept of meditation. You should look for a balance between moments of action or experimentation and moments of reflection.

It is my experience that I work faster and deliver better work, if I keep my mind clear rather than when I just work long hours. Sometimes I have worked extremely late, trying to crack a problem and then at some point given up and gone to bed. Then when I have woken up the next day, one of two things has happened. One: while standing in the shower the solution to my problem has suddenly popped into my mind. Then during the car ride to work, the plan has fallen into place in all its details. Two: I would look at my work from the previous night and discover that it was full of mistakes I would never normally make. On some occasions, for instance, I have discovered that I accidentally did not save my file correctly so losing a lot of work. If you keep your mind clear, you make fewer of these mistakes and you get more spontaneous creative insights.

I now relax more, but that doesn't mean I don't try hard. The 'trying hard' now focuses on the discipline of keeping my priorities straight. It is a continuous focus for working on the right things and doing them with full attention. It also doesn't mean that I never work long hours; I just don't do it on a structural basis. If I work long hours, it is mostly because I want to give a project that extra push to take it to the next phase. There is a thin line there though: a few years ago, I would always find reasons to make that small sprint to speed up a project, then before I knew it, I was making small sprints continuously. If I work late because a situation forces me to do so, I now conclude that I misjudged the planning or workload. I will make sure things get done anyway. Then I will try to learn from the mistake to make sure it doesn't happen again. You could even say that *refraining* from working long hours, requires more discipline than *working* them.

So 'trying hard' is not the same as 'running like hell'. If you constantly run like hell, you can't be persistent in the long-term, for you *will* wear yourself out at *some* point. 'Trying hard' is keeping your eye on the ball when it matters. As described in the previous section about Karate, your path for self-development can be considered a marathon that can only be won through self-discipline, hard training and your own creative efforts. Use all the means available to you in reaching your objectives, so you know your efforts were never the limiting factor. In the context of battle, Samurai Musashi writes: "This is the truth: when you sacrifice your life, you must make the fullest use of your weaponry. It is false not to do so and to die with a weapon yet undrawn."

Take care of yourself

"Take care of yourself" is probably the wisest thing my teacher ever taught me. He said it to me every time I left the dojo after training. I only reached the point where I truly trusted my ability to protect and take care of myself a few years after he passed away. I needed time to trust myself in saying "no" to influences that reduced my quality of life. Excessive workloads and tricky politics are examples of these influences. In sport, those negative influences were overtraining and constantly criticizing myself for not being strong enough. I more or less needed a permanent injury to stop me from doing those things.

In work, luckily, I have learned to change these habits. I take care of myself a lot better than I used to. As I maintain this, negative influences are gradually fading away. This happens simply by refusing to put energy into the negative and instead focusing on building great work with great people.

I mentioned earlier that in my time as competitive fighter in Judo, I most likely would have performed better if I had trained less. You need to respect the limits of both your body and mind. If either of those gives up on you, you are worth nothing as a professional. It is equally important in sports and business to train, eat, sleep and relax well. Things you do, need to feel light-footed. If they don't, you should reflect on what causes that feeling and see if it can be changed. If it can't be changed, you can either accept it or walk away.

Keeping your mind clear is dependent on keeping your body healthy.

Your mind can only be clear and relaxed, if you have the right energy levels in your body. If some illness is creeping in I can feel it immediately in my focus. Keeping focus then costs more energy, which wears me down during the day. Regular small breaks, getting to bed early and sometimes working at home for a day are things that generally help in preventing sickness. When my body is healthy, keeping focus does not cost me energy. I feel as fit at the end of the day, as I feel in the morning. When something stressful happens, I can feel this change. The focus starts costing effort again. The sooner I detect this, the sooner I can act and have a clear mind again.

Don't forget to enjoy the things you do. By the time I had a permanent injury and could not practice Judo anymore, I discovered that there had definitely been times when I had forgotten to enjoy the privilege of doing Judo, of moving freely and explosively with my body. I was often too busy becoming stronger and better to stop and enjoy my abilities.

In the end, it comes down to the fact that doing anything to extremes is not beneficial. Eating healthily is good, but that does not mean extreme diets. Doing sports is good, but that doesn't mean you need to run a marathon each week. This may sound boring, but the sequence of small events is what makes a balanced attitude powerful. If you are balanced, people will like working with you, they will be happy for you if you get promoted, the quality of your work will be better, which means you will get to work on the nicer projects and in the nicer jobs and companies. Your projects will be of better quality and there will be fewer problems that cannot be overcome. The atmosphere around your projects will be a good environment to work in. A continuous flow of good stuff will move into your life, including the right friends, colleagues and the right partner for life. The positive energy coming from that is not the same as the thrill that comes from doing extreme sports, from going out late and partying wildly or from being in a competition and knocking a strong opponent out, but altogether it is still the true and positive energy that makes life worth living. With this energy, you will see more opportunities for unexpected things that can enrich your life. You will be stronger if painful things do happen to you, like serious illness or the death of a loved one. You can more easily decide to take a month off to make that trip that you've always wanted to make. You can decide to take a leap of faith and move into that job you've always wanted or start that company you've always dreamed about. Even if your attempts fail or turn out less well than you thought they would, you will know you gave it the shot it deserved. This will always help you in making the next good decision in your life.

The Other Side

One day, a young Buddhist monk on his journey home came to the banks of a wide river. Staring hopelessly at the great obstacle in front of him, he pondered for hours on how to cross such a wide barrier.

Just as he was about to give up his journey he saw a great teacher on the other side of the river.

The young monk yelled over to the teacher:
"Oh wise one, can you tell me how to get to the other side of this river?"

The teacher pondered for a moment looked up and down the river and yelled back:
"My son, you are on the other side."

Be grateful, not arrogant

As you progress in your professional career, you are likely to gradually become more successful. For some people arrogance waits around the corner. Arrogance blocks your ability to grow and having too much of it will take the heart out of your success. It is important to remain grateful for the successes that come your way. Success can fade at any moment. I still remember how invincible I felt when my body was fully trained and healthy. In one single second, one small piece of my knee broke and took away all my fighting skills forever. I have learned to appreciate my health, skills, success and even the fact that I have a job and income at all. If we are successful, we sometimes assume that it will continue in the future.

Be grateful for the talents you have and for the experiences you gain in your quest for self-development. It is good if you are conscious of your skills and qualities and therefore feel that you deserve a certain promotion or increase in salary, but don't forget to be grateful if you actually get them. If you can feel gratitude in parallel to pride, you will take the edge off arrogance. Celebrate successes, but stay focused on the next steps.

Wanting promotion and to grow too fast may distract you from the things you are working on now. Sometimes you are already on the right track but just need a bit more time. Many jobs or companies look better from a distance. Switching jobs may sometimes feel like an easy way of growing faster, especially if you are approached by head-hunters who flatter you and boost your view of your own value. However, every company and job has disadvantages that may be hard to see from the outside. If you feel another job or company is better than the one you have now, think hard about how you will feel in your new job after you have worked there for at least two years. Think what the new job will add to your long-term objectives and think about your current job in relation to your long-term objectives. In the context of the previous story of the two monks at the river: it is easier to assess whether the grass is really greener on the other side, if you fully appreciate what you have now.

You are as strong as the people supporting you

In many ways, you are dependent on the people you work with and on the people in your private life, so choose them wisely. You will always need the support of a variety of people for a variety of purposes. Support can vary from knowledge to friendship and from critical feedback to a pat on the shoulder every now and then.

As a fighter you also need many kinds of people to support you. First of all you need a good teacher, you need sparring partners, a mental and strategic coach, maybe a specialized trainer for power training, specialized guidance in nutrition and rehabilitation after injury and last but not least you need opponents. In some cases, one person may fulfill multiple roles. A teacher for instance, can also be your coach in competition. You don't necessarily need to find everything in one person though.

This is also true in business. It is nice if you have a manager who is also a mentor and can help you in specialized knowledge concerning your profession. However, you rarely find all of this in one person. If you do, consider yourself lucky. The people I have seen grow fastest in their careers are the ones that were managed by a mentor: a manager who would help them learn and grow, help them to work on the right projects and to get the right promotions. In some cases those mentors even helped them in finding their next job and mentored them into their new role in another organization.

Even though it is a blessing to have a manager who can coach you in all aspects of your job, it is also a risk. We all have our own limited scope and way of looking at the world. If you learn all the things you know from only one person, you risk not broadening your own scope enough. So even if you have the perfect manager who has a broad vision, true depth of knowledge and willingness to share it with you, keep looking around for additional people who can teach you the things you don't know yet.

Try to visualize the kinds of guidance you need in your job. Evaluate how much of that your manager can and is willing to offer. Maybe your manager knows a lot about your field of work, but is not really a good coach or it could be the other way around. For each piece of the puzzle, try to find multiple people who can support you. Make sure you offer them something in return by sharing your knowledge, your efforts or maybe just your friendship. You may have a lot more to offer than you think.

In my job as researcher in the field of online and offline marketing, I touch on a large amount of specialized knowledge. It would be impossible for me to oversee all of it. If I start working in a new environment, I look for the people who really know what they are doing and who are eager to move forward and develop their knowledge. I try to build professional relationships based on exchange of knowledge and a mutual interest in doing great work.

Some people are strong in knowing the organizational structure or parts of it. They can help me connect to the right people for certain topics. Others are specialists in finding the right data sources.

Others might be people I trust to talk to when I have personal things on my mind. I like sparring about new ideas with people who will offer constructive criticism. Without this network of people around me, my work would not only be harder, it would be a lot less fulfilling. Find people who can enrich your work and working experience. Make sure they get at least as much value out of the cooperation as you.

Your network can be built both inside and outside your organization. When I was a consultant in marketing and advertising research, I learned most from the knowledge exchange with clients. People don't necessarily need to be a permanent part of your network. In a way, everyone can be your teacher. You could learn something valuable from a person you meet only once in a conference, in a meeting or even sitting next to you on the bus. Everyone has at least one unique skill. Look at people and at what qualities they have, ask them about their qualities and skills and experiment with whether they work for you.

Someone coming from a different angle can shed surprising light on matters that concern your job or about who you are as a professional. This was one of the reasons for me writing this book about Samurai and business. Looking at the experience of a fighter can be refreshing and may bring new insights. Talking to a broad variety of people prevents you from being trapped into the loops of thinking that are common in your field of work.

Don't overlook the *care takers*. These might be the receptionist, secretary, cleaning person, postman, IT-guy, etcetera. These people are the backbone of an organization and there is nothing more annoying

for them than being ignored by someone for months and then suddenly being chased when they are needed. Make sure you understand their roles and how they need to be informed by you or how you can make it easy for them to process your requests. Give them all the information they need in a timely way.

Finally, don't forget to value your competitors. A fighter cannot develop without opponents. This is also true for businesses. Let yourself be inspired and challenged by your competitors. Give them credit for the things they did well. You are competing, but you are also colleagues. After his victory against his strongest opponent Sasaki Kojirō, *The Demon of the Western Provinces*, Samurai Musashi took a moment to reflect and thank his opponent for being strong and having challenged his skills. With a wave of admiration and respect, he realized he would never meet an opponent like him again.

Look around you with an open mind, with eagerness to learn and willingness to share your knowledge. People need to like working with you. They need to experience your contributions as valuable and your professional attitude as reliable. If that is true, you will have an army of people supporting you in everything you do.

Judo

Judo is probably the most well known martial art globally. It is an Olympic sport, but was originally not intended for competition. Judo was founded in Japan by Jigoro Kano. Kano was born into a relatively affluent family. He had an academic upbringing and from the age of seven he studied English, Japanese calligraphy and the Four Confucian Texts (Chinese ethical and philosophical system regarding the cultivation of human beings). Being bullied at school at a young age was the catalyst that caused Kano to start practicing Jiu-jitsu. In 1882, Kano founded his first Judo school, Kodokan.

In modern Judo the objective is to throw your opponent on the back and in ground fighting to hold him there for a period of time. You can also make the opponent surrender using an arm-lock or choke. The maximum score for a throw or the surrender of the opponent is *Ippon*. *Ippon* immediately ends the fight, so a Judo fight can be lost or won in any second. Ippon symbolizes death of a warrior who is taken by surprise.

Kano taught two Judo principles. The first is *Seiryoku Zen'yō* (maximum efficiency, minimum effort): resisting a more powerful opponent will result in your defeat, while adjusting to and evading your opponent's attack will unbalance him. This will reduce his power, so weaker opponents can beat significantly stronger ones. The second principle is *Jita Kyōei* (mutual welfare and benefit): Kano was convinced that practice of Judo, while conforming to his ideals, was a route to self-improvement and the betterment of society in general.

Mutial welfare and benefit

I believe that anyone living by honorable principles has the power to change his or her environment to a certain extent. Each person you change, changes other people as well. So each individual can invoke a powerful chain reaction that can be either positive or negative. I believe Jigoro Kano was a wise man in linking his principles of self-development to those of the betterment of society, as described in the previous section about Judo. This is exactly the point where the egoistic nature of people meets the common good. Personal success does not need to conflict with the success of others if your work is based on the right principles. Financially successful companies can create value by offering great products for a fair price and by providing people jobs at a fair salary. They will remain valuable as long as they focus on serving the evolving needs of end users and clients. If you establish that focus in cooperation with your colleagues, while mutually sharing knowledge and credit, colleagues can grow along with you and you will inspire each other to deliver great work.

Self-development adds power to integrity and allows you to confront people or difficult situations without aggression or unnecessary harm. If you embrace self-development as a continuous process, you are likely to surprise yourself positively. In the long-term, your development process then becomes an endless chain reaction that can take you to points you never anticipated. Then when you are successful, the will to develop yourself is there to prevent you from becoming arrogant, since there is always more to learn.

I want the office I work in to be like a dojo: a place where experiments lead to valuable progress and where people challenge each other sharply but respectfully. I don't expect it will ever be perfect, but I believe I can contribute to supporting the right work environment, just as anyone else can. You will always encounter people with objectives that conflict with yours. If you do, connect with them with true respect and feel what drives them. You might find a way to unite and if not, be confident you can defend your position. If you prepare well and invest effort to the best of your ability you cannot lose. Take responsibility for the good and bad things that happen to you. Improve without judging yourself (or others) for not being perfect. You always gain experience regardless of outcome.

Train your internal and external awareness to stay clear under any kind of pressure. Align your objectives and actions proactively to your environment (center control) and manage expectations (distance control), while remaining healthy and relaxed (balance). Learn to apply these principles with totality and a hundred percent consistency utilizing self-discipline, patience and creativity. Look fear in the eyes, while still doing what you think is right and necessary. Take joy in the things you do and accept the annoying things that are inevitably connected to it. Celebrate successes together with colleagues, friends and loved ones. Do so with pride and gratitude.

I sometimes wonder if I would have done things differently had I known the price of my injury. I then realize that without this path I have chosen, I would also never have met my teacher and friend Jan.

Knowing this trade-off in advance, the choice would have been an impossible one. It is probably a good thing that sometimes we don't know how much we will pay or how much we will get in return. The human mind is not capable of weighing such options, for some experiences are just priceless and some losses can only be accepted if they are unavoidable.

This is how I see things. This is how I do things. I wouldn't feel happy doing them any other way. I hope you find this view refreshing and that it helps you in being successful and enjoying your profession. Any smile you evoke, any person you inspire and any great idea or product you contribute to, are all pieces of your own Samurai sword: they cannot be stolen, they can only be earned. I hope you will help me in living this message and sharing it, so we can all enjoy our work more and serve our clients to the best of our ability.

Thera Benjaminsen

Thera Benjaminsen is a Dutch art painter born in the early 1950s in Nijmegen, the Netherlands. As a young girl she played the violin and practiced ballet.

In the mid-1970s she and her daughter migrated to Italy, where she met her Norwegian husband-to-be. They married in Germany and moved to communist Russia. A year later, they moved to Iran where her second daughter was born. After experiencing a revolution and the first years of the Iran-Iraq war, Thera and her family settled in Oostvoorne, in the Netherlands in the early 1980s, where she still lives.

As an adult, Thera discovered a natural talent for drawing. In the mid-1990s, drawing became painting and soon after she developed her sculpting skills. In 2005, her passion for painting motivated her to further develop her skills in model painting courses at the Academy of Arts in Rotterdam. In doing so, she continued her quest in personal development, always working towards freedom of expression.

Her colorfully expressive and characteristic paintings have expanded into a large intriguing personal art collection.

156

Text: Joris Merks
Artwork and cover: Thera Benjaminsen
Editor: Dawn Sackett
ISBN 978-0-929652-21-4
LCCN 2012946314

Meghan-Kiffer Press
Tampa, Florida, USA
Innovation at the Intersection of Business and Technology
www.mkpress.com

www.samurai-business.com
www.facebook.com/samuraibusiness
plus.google.com/samuraibusiness

Special thanks to my proof readers:
Claire Benjaminsen, Paul van Eden, Allan Hall, Tom Wyvill, Nick Arini,
Henry Eccles, Brian Riggs, Mark Jansen, Rick Malins, Tony Fagan and
Amanda Welsh.

Also available from Meghan-Kiffer Press

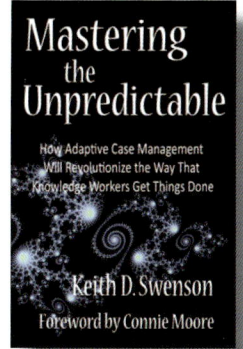

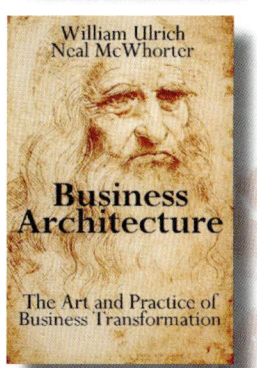

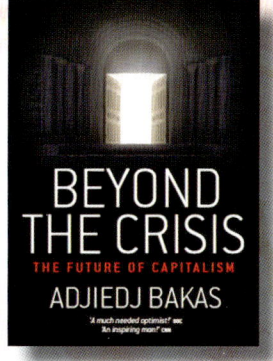

www.mkpress.com

Samurai principles for success based on integrity

- *Dedicate yourself to a purpose beyond power, control or earning money.*

- *Develop yourself to the benefit of the world around you.*

- *If you encounter a problem: change it, accept it or leave it.*

- *Stay connected to yourself and the environment under any kind of pressure.*

- *Take a close view of distant things and a distant view of close things.*

- *Balance careful planning with creative and flexible execution.*

- *Don't fight inevitable developments.*

- *Be respectful, yet clear and sharp.*

- *Reflect without judging.*

- *Look fear in the eyes while doing what you think is right and necessary.*

- *Inspire people and celebrate successes with gratitude, not arrogance.*

- *Be helpful and generous, yet choose the people around you wisely.*

- *Take care of yourself and those around you.*